Microelectronic Systems
LEVEL 2

IAN SINCLAIR, BSc, M.I.E.E.

Lecturer in Electronics,
Braintree College of Further Education

HOLT, RINEHART AND WINSTON
LONDON · NEW YORK · SYDNEY · TORONTO

Holt, Rinehart and Winston Ltd: 1 St Anne's Road,
Eastbourne, East Sussex BN21 3UN

British Library Cataloguing in Publication Data

Sinclair, Ian
  Microelectronic systems Level 2.
  1. Microprocessors
  I. Title
  621.3819′5835   TK7895

ISBN 0–03–910373–0

Printed in Great Britain by Mackays of Chatham Ltd, Chatham, Kent

Last digit is print number: 9 8 7 6 5 4 3 2 1

Contents

PREFACE   v

1 Numbers   1

*Binary Numbers 1; The Byte 2; Hexadecimal and Octal 4; Summary 1.1
5; Exercises 1.1 8; Binary Addition 8; Negative Numbers 10; Summary
1.2 11; Exercises 1.2 11; Two's Complement 11; Binary Subtraction 12;
Other Functions 13; Summary 1.3 14; Exercises 1.3 14; End-of-Chapter
Test 15*

2 Inside the Microprocessor   16

*Registers 16; Loading and Storing 17; Summary 2.1 19; Exercises 2.1 19;
Byte Manipulation 19; Shift and Rotate 22; Instructions—Operator and
Operand 24; Summary 2.2 24; Exercises 2.2 25; Extended Block Diagram
25; The Fetch-Execute Cycle 29; Summary 2.3 30; Exercises 2.3 31;
End-of-Chapter Test 31*

3 Creating Programs   33

*Algorithms 33; Algorithm to Flowchart 34; Summary 3.1 38; Exercises
3.1 39; Deriving a Flowchart 40; An Example of Two-Byte Addition 43;
Summary 3.2 44; Exercises 3.2 44; Tracing the Action 46; The ROM 48;
Summary 3.3 50; Exercises 3.3 50; End-of-Chapter Test 51*

4 Memory Systems   52

*Memory 52; Memory Organization 54; Summary 4.1 56; Exercises 4.1 56;
Memory Selection: Address Decoding 57; Chip Select/Enable 58;
Summary 4.2 60; Exercises 4.2 61; Read/Write Control 61; Data in
Memory 62; Summary 4.3 63; Exercises 4.3 64; End-of-Chapter Test 64*

5 Instructions 66

Instruction Sets 66; Instruction Groups 66; Summary 5.1 68; Exercises 5.1 69; Arithmetic and Logic 69; Test and Branch 72; Summary 5.2 73; Exercises 5.2 74; Addressing Methods 74; Immediate Addressing 75; Extended Addressing 76; Indirect Addressing 78; Summary 5.3 79; Exercises 5.3 80; End-of-Chapter Test 80

6 Looping Programs 81

Branching and Looping 81; Summary 6.1 84; Exercises 6.1 86; The Branch Step 86; Summary 6.2 90; Exercises 6.2 91; An Adding Program 92; Summary 6.3 98; Exercises 6.3 99; End-of-Chapter Test 99

7 Buses and Interfaces 101

Buses 101; Summary 7.1 104; Exercises 7.1 104; Interfacing 104; Summary 7.2 108; Exercises 7.2 109; Code Conversion 109; Voltage Conversion 110; Timing Conversion 110; Summary 7.3 112; Exercises 7.3 113; End-of-Chapter Test 113

Appendix A 114

SI-MPU-2 Program

Appendix B 118

The MENTA Assessment Unit

INDEX 120

Preface

The TEC Level 2 Microelectronic Systems unit is intended as a follow-on to the U79/602 Level 1 half-unit, which is the introductory course for Microelectronic systems. To allow for the usual gap of several weeks between the end of the Level 1 course and the start of Level 2, this textbook will cover many of the basic principles which were dealt with in more detail in Level 1. This approach also permits students who have obtained an exemption from the Level 1 unit to proceed with study at Level 2. At this level, however, the subjects are treated in greater depth, and with less use of long explanations than was considered necessary in the lower-level book.

As was noted in the Level 1 book, nothing is an adequate substitute for practical experience with microprocessor systems, and students should make every effort, no matter what frustrations are encountered, to come to grips with a microprocessor assessment unit for as long and as often as is possible. The MENTA unit described in Appendix B is a more recent design which has considerable advantages over older units.

1

Numbers

Binary Numbers

THE normal scale of numbers which we use for counting is the scale of ten, or *denary* scale, which uses digits represented by the symbols 0, 1, 2, 3...8, 9. Numbers are written as groups of digits, and the position of a digit in a group indicates its *significance* — whether it represents units, tens, hundreds, thousands or any other power of ten. For example, the number 763 means seven hundreds, six tens and three units.

The easiest way to deal with numbers electronically is to have only two digits, '0' and '1'. '0' can be represented by zero voltage or zero current, '1' by some value of voltage or current which is not zero, and numbers which use only these two digits are called *binary* numbers. Like a denary number, the significance of any digit in a binary number is indicated by its position in the number, and each position in a binary number is valued at twice as much as the position immediately to the right. For example, the binary number 101101 represents, reading from right to left, a 1, no 2s, one 4, one 8, no 16s and one 32, making a total of 45 in denary when we add them all up. Any whole number that can be written in denary can also be written as a binary number, but the binary number will contain more digits, and the digits will all be either zeros or ones.

The binary code is not the only way of arranging the two digits 0 and 1 so as to represent numbers. Two other systems, Gray code and excess-three code are illustrated in Fig. 1.1; they have the advantage over binary in that only one digit in the number is changed when the

Denary	Gray Code	Excess-3 Code	
0	0000		0011
1	0001		0100
2	0011		0101
3	0010		0110
4	0110		0111
5	0111		1000
6	0101		1001
7	0100		1010
8	1100		1011
9	1101		1100
10	1111	0100	0011
11	1110	0100	0100
12	1010	0100	0101
13	1011	0100	0110
14	1001	0100	0111

Figure 1.1 *Gray code and excess-3 code. The excess-3 code is found by adding 3 to each denary digit, then converting each digit to 4-bit binary. The Gray code cannot be found by a simple conversion, and Gray-code arithmetic has to be performed by first converting Gray code to binary.*

count changes (up or down) by one. Neither of these codes is used within microprocessor systems, however, but they can be encountered in machine-control systems, particularly as methods of encoding the position of a shaft.

Microprocessor systems make much use of binary numbers that are arranged with the digits in groups. A binary digit is called a *bit* (which is simply a shortened version of *bi*nary dig*it*), and a group of bits used in a microprocessor is called a *byte.* Many microprocessors use a byte of eight bits, so that we shall refer to a byte throughout this book as meaning eight bits. Microprocessors which use 16-bit groups are being manufactured but are not in common use. When these devices appear, the contents of this book will still be relevant, but the word 'byte' will have to be interpreted as meaning 16 bits for these microprocessors.

The Byte

A byte, then, for an 8-bit microprocessor, means a group of eight binary digits, and the number quantity that the byte represents is decided by

(a) *Denary-to-binary*

Write down the denary number, and divide it by 2. Write the whole part of the result underneath, and any remainder (0 or 1) alongside. Now divide the answer by 2 again, and keep doing this until only a 1 remains. Now read the remainders in order from the bottom upward – this is the binary number.

Example

```
2 ) 176
2 )  88   0
2 )  44   0
2 )  22   0
2 )  11   0
2 )   5   1
2 )   2   1
2 )   1   0
          1
```

176 denary = 10110000 binary

(b) *Binary-to-denary*

Write down the binary number. For each digit in the number, write down the power of two if there is a 1 in that place. The power of two can be found by doubling each number as you move to the left. Then add up all the denary numbers. This is the denary equivalent. In the example, powers which are bracketed are ignored in the total because there was a 0 in that place.

Example

01100111

```
    1
    2
    4
   (8)
  (16)
   32
   64
 (128)
```

01100111 binary = 176 denary

Figure 1.2 *Converting between binary and denary scales.*

the pattern of the bits in the byte. For example, 00000001 and 10000000 are both patterns which contain seven zeros and a one, but their arrangement is quite different. The difference is emphasized when we find the denary equivalent of each. 00000001 is denary 1, and 10000000 is denary 128 (though see later for negative numbers). Each different pattern of bits in a byte represents a different denary value, so that the denary number is quite a convenient way of expressing the value of a byte, and one which uses fewer digits. Fig. 1.2 shows methods of converting binary values into denary and denary to binary.

Denary scale is not ideal, however, as a method of writing down the value of a byte. The base number of the denary scale is ten, and ten is not a power of two. 2 to the power of 3 is 8, and 2 to the power of 4 is 16, so that base numbers of 8 or 16 would logically give us more useful representations of bytes.

Hexadecimal and Octal

The hexadecimal scale, always abbreviated to hex, a scale whose base is 16, is used extensively as a convenient way of representing binary numbers. Because we have no symbols to represent digits of values 10 to 15 in denary, we use the letters A to F for these purposes, so that a hex count from zero to 16 (denary) looks as shown in Fig. 1.3, with denary and binary equivalents for comparison. Note that we usually follow a hex number with an 'H' to distinguish it from other types of numbers. Each digit of a hex number represents four bits of the binary number, so that conversion between hex and binary (Fig. 1.4) is quick and easy when the binary number bits are arranged into bytes. All that is needed is a good knowledge of the equivalent hex digit for each binary number between 0000 and 1111 (binary), 0 to F (hex). Conversion between hex and denary is carried out in the same way as conversion between binary and denary, but using 16 (denary) as the base number rather than 2 (Fig. 1.5).

The octal scale (scale of eight) has the advantage of not needing any additional digits, because it uses the digits 0 to 8 only, but an octal representation of a binary byte does not correspond so neatly as the hex version, and is not so easy to convert. A number such as 0010 in binary

Denary	Hex	Binary
0	00	0000
1	01	0001
2	02	0010
3	03	0011
4	04	0100
5	05	0101
6	06	0110
7	07	0111
8	08	1000
9	09	1001
10	0A	1010
11	0B	1011
12	0C	1100
13	0D	1101
14	0E	1110
15	0F	1111
16	10	0001 0000

Figure 1.3 *Denary, hexadecimal and binary numbers.*

(a) *Hex to binary.* Write down, using the table of Fig. 1.3, the binary equivalent of each hex digit in the number, reading from right (least significant) to left (most significant).

Examples: 3EH ... binary 3H = 0011, binary EH is 1110, so 3EH is 11100011.

1F2AH ... converts to 0001111100101010.

(b) *Binary to hex.* Group the binary number into sets of four digits, starting at the right-hand side (least significant digits). Convert each group of four (and any final smaller group) into hex, using Fig. 1.3.

Examples: 1011100110110 groups as 1   0111   0011   0110

and this converts to   1   7   3   6

which is the hex number 1736H

1101100001101100 groups as 1101 1000 0110 1100

and this converts to   D   8   6   C

which is the hex number D86CH.

Figure 1.4 *Converting between binary and hex scales.*

is 02 in octal, but 1010 in binary is 12 in octal, and 1111 (binary) is 17 in octal. A 4-bit binary number may need either one or two octal digits, whereas the same binary number, written in hex, will require only one digit. Conversions between binary and octal are therefore less easy than conversions between binary and hex, and this has resulted in the octal scale being used to a much lesser extent than it used to be (I personally have never seen a microprocessor program written in octal). The methods of converting between denary and octal are shown in Fig. 1.6.

Microprocessor systems use hex notation almost universally, so that this scale is by far the more important as a method for working with patterns of binary digits. A byte of eight bits is represented as a two-digit hex number, a sixteen-bit address as four hex digits. Remember, however, that the microprocessor system always makes use of binary digits in terms of electrical signals, and hex code is simply a convenient way of writing these numbers.

Summary 1.1

Numbers are most easily represented by electrical signals if a binary number scale is used. The conventional binary scale uses two digits, 0

(a) *Hex to denary.* Write down the denary value of the least-significant hex digit (right-hand side). Now write underneath this the value of 16 × the next digit. If there are more digits, write down the value of 256 × the next digit, then 4096 × the next digit. Finally, add up these numbers.

Examples: Convert 3AH to denary:

$$A = 10$$
$$3 \times 16 = 48$$
$$3AH = 58$$

Convert 7C2D to denary:

$$D = 13$$
$$2 \times 16 = 32$$
$$12 \times 256 = 3072$$
$$7 \times 4096 = 28\,672$$
$$7C2D = 31\,789$$

(b) *Denary to hex.* Write down the denary number. Divide it by 16, writing the whole number result underneath, and any remainder *in hex* to the right. Continue until the last result is less than 16, then read the remainders from the bottom up, including the last result, which is the first digit of the hex number.

Examples: Convert 184 to hex:

```
16 )  184
16 )   11      8
        0     0B
      184 = 0B8H
```

Convert 27 126 to hex:

```
16 )27 126
16 )  1695     6
16 )   105     F
        6      9
   27 126 = 69F6H
```

Figure 1.5 *Converting between hex and denary scales.*

and 1, with the position of each 1 in the binary number indicating its relative value. Each digit, 0 or 1, is referred to as a bit (binary digit), and a group of eight bits is usually referred to as a byte. Because of the number of digits in a binary number, and the ease with which one large binary number can be confused with another one, hex (hexadecimal code) is normally used for writing numbers. One hex digit can represent four bits, because the hex scale uses number and letter symbols for denary numbers up to 15, which is 1111 (binary). The hex number is usually written with an 'H' following it.

Denary	Octal	Binary
0	000	0000
1	001	0001
2	002	0010
3	003	0011
4	004	0100
5	005	0101
6	006	0110
7	007	0111
8	010	1000
9	011	1001
10	012	1010
11	013	1011
12	014	1100
13	015	1101
14	016	1110
15	017	1111
16	020	0001 0000

Conversions

(a) *Binary and denary to octal.* To convert binary to octal, group binary number in threes, starting with least-significant digit (right-hand side). Write down the octal equivalent for each group, using the table above. To convert denary to octal, divide repeatedly by eight, and read remainders from bottom up, as before.

Examples: Convert 011001110 to octal:

$$\text{Group as } 011 \quad 001 \quad 110$$
$$\text{and convert as } 3 \quad 1 \quad 6 = 316 \text{ (octal)}$$

Convert 1265 to octal:

```
8 ) 1265
8 )  158   1
8 )   19   6
       2   3
```

1265 denary = 2361 octal

(b) *Octal to binary and denary.* To convert octal to binary, write down a group of three binary digits for each octal digit, using the table above (start either end). To convert octal to denary, write down the least-significant digit (right-hand side), then 8 × the next digit, then 64 × the next digit, then 512 × the next and so on (each multiplier is eight times the previous one).

Examples: Convert 172 octal to binary:
Write down the groups: 001 111 010, giving 001111010 binary.
Convert 2516 octal to denary:
Write down the denary numbers:

$$6$$
$$1 \times 8 = 8$$
$$5 \times 64 = 320$$
$$2 \times 512 = 1024$$
$$\text{And adding} = 1358 \text{ denary}$$

Figure 1.6 *Octal, binary and denary number scales and conversions.*

Exercises 1.1

1. What is meant by a denary scale?
2. What is meant by a binary scale?
3. In the denary number 954, what does the 9 mean?
4. Why is a binary scale so well suited to electrical representation of numbers?
5. What is a bit?
6. What is conventionally meant by a byte?
7. Convert the following to denary: (a) 10110110 (b) 01101101.
8. Convert the following denary numbers to 8-bit binary: (a) 187 (b) 092.
9. Convert the following bytes into hex: (a) 01101011 (b) 10101101.
10. Convert the following hex numbers to binary: (a) 2BH (b) F6H.

Binary Addition

The rules of binary addition are summarized in Fig. 1.7. An addition of two 1's generates a 0, with 1 carried to the next higher place; an addition of three 1's gives a 1, with another 1 carried. The carry is a bit which must be added in to the addition of the next two bits in the next higher place in the numbers, just like the carry in denary arithmetic.

For example, the addition of 1011 to 0110 (Fig. 1.8) gives 10001. Starting with the least significant digits (on the right-hand side), the

$$0 + 0 = 0 \qquad 1 + 1 = 0 \text{ and carry } 1$$
$$0 + 1 = 1 \qquad 1 + 1 + 1 = 1 \text{ and carry } 1$$
$$1 + 0 = 1$$

Figure 1.7 *Rules of binary addition.*

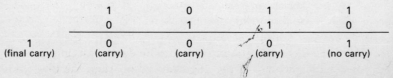

$$1011 + 0110 = 10001$$

Spread out, this looks as follows:

	1	0	1	1
	0	1	1	0
1	0	0	0	1
(final carry)	(carry)	(carry)	(carry)	(no carry)

Figure 1.8 *A sample addition, showing the carry bits.*

Binary	Hex
11011001	D9H
10110111	B7H
10010000	190H

1 (carry)

Figure 1.9 *How addition of two single bytes can produce an answer which is larger than one byte.*

addition of 1 to 0 gives 1, which is written down as the least-significant bit of the answer. Adding the next bits, 1 + 1, with no carry, gives 10, so that the 0 is written down, and the 1 carried. The next addition is 1 + 0 + 1 (carry), giving 10 again, so that 0 is written down and 1 carried again. The last addition is 1 + 0 + 1 (carry), so that 0 is written down, but since there are no more additions to perform, the final carry is written down as the left-hand side (most-significant place) of the number. This gives the answer 10001.

In such a binary addition, a carry bit from the addition of the most-significant (left-hand) bits will cause the answer to be one bit larger than the original numbers, as illustrated in the previous example. This is simple enough when the addition is performed on paper, but not quite so simple when the addition is done electrically in an 8-bit microprocessor store. When the numbers that are added do not fill the store, there is no problem, but when the two eight-bit numbers generate a carry out of the most significant bit when they are added (Fig. 1.9), then that carry bit cannot be stored in the same store as the rest of the byte, and the number that is left there is not the correct result of the addition. For example, the addition of 10101100 to 11000101 gives on paper 101110001, and since the carry cannot be stored along with the other eight bits, the number which remains after addition is 01110001, which is certainly not the correct answer to the addition — try converting these numbers to denary to see the difference that the absence of the carry makes to the value.

The microprocessor caters for this problem by keeping one part of a special store, called the *status* (or *flag*) *register*, just for a carry bit which will be generated when the addition of two one byte numbers causes an answer greater than one byte. This carry bit can then be used if the answer has to be displayed, or it can be added in to the least-significant place of another addition if needed. For example, if two 16-bit numbers are to be added (Fig. 1.10), and they can be dealt with only in 8-bit units because of the capacity of the microprocessor, then the lower 8 bits of each can be added first. If this addition generates a carry, that carried

Binary		Hex
01011011	10110111	5BB7H
00010011	11001010	13CAH
01101111	10000001	6F81H

(carry)

Figure 1.10 *The addition of two bytes may need a carry transferred from the lower byte to the higher one. This carry must be stored, because the additions are done in sequence.*

bit must then be added in to the lowest bits of the next group of eight, just as it would have been added if the microprocessor had been able to deal with 16 bits in one addition.

Negative Numbers

We represent denary negative numbers on paper by using the negative sign ($-$), and we can think of subtraction as the process of adding a negative number, so that $5 + (-3)$ has the same meaning as $5 - 3$, and is equal to -2 in denary. Microprocessors make no provision for either positive or negative signs, so that some other method of indicating when a number is negative has to be used. The method that is universally employed is to use the most-significant bit (left-hand side) of the binary number as a sign bit. A signed 8-bit binary number, therefore, consists of seven bits of number digits and one sign bit, and the convention is that a 1 in the sign-bit position means that the other seven bits represent a negative number, a 0 as a sign bit means that the other seven bits represent a positive number. Using this convention, 8-bit numbers from 00000000 to 01111111 are regarded as positive (denary 0 to 127), and numbers from 10000000 to 11111111 (denary -128 to -1) are regarded as negative.

This convention might appear at first to cause difficulties, because a number such as 10001001 could be taken either as a positive number, equal to $128 + 8 + 1 = 137$ (denary), or as a negative number with quite a different value. The microprocessor, however, is programmed to treat each byte in the same way, no matter whether we think of it as being negative 7-bit, positive 7-bit, or positive 8-bit, and it is *entirely* up to the programmer to decide how a number shall be interpreted. This is possible because a binary negative number is a *two's complement* of a positive number.

Summary 1.2

Binary numbers can be added in the same way as denary numbers, remembering that $1 + 1 = 10$ in binary. As in denary arithmetic, when a 1 has to be carried, it is then added in to the addition of the next bits (moving from right to left). The bit on the right-hand side is the least-significant bit (l.s.b.), that on the left-hand side is the most-significant bit (m.s.b.).

Negative signs, if needed, can be represented by using the m.s.b. as a sign indicator. The convention is that the m.s.b. is 1 if the number is negative, 0 if the number is positive. When this scheme is used for single-byte numbers, only seven bits are available to use for the number itself. The programmer must decide whether an 8-bit number is to be treated as signed (7 bits of number) or unsigned (8 bits of number).

Exercises 1.2

1. Write down the rules of binary addition.
2. What is a 'carry bit'?
3. Add: (a) 0110 to 1011 (b) 10011 to 11000 (c) 10011001 to 00011011.
4. How is a carry from the m.s.b. of an addition treated by a microprocessor?
5. What is the status register?
6. When is the carry bit of the status register set to 1?
7. How is this carry bit used?
8. How is a negative number represented?
9. How might the method of representing negative numbers cause confusion?
10. How does the microprocessor treat negative numbers?

Two's Complement

The *complement* of a binary number is formed by exchanging each 1 in the number for a 0, and each 0 for a 1. For example, the complement of 1001 is 0110. The *2's complement* of an 8-bit number is formed by adding 1 (00000001 in 8-bit binary) to the complement, so that the 2's complement of 00010101 is 11101010 + 1 = 11101011. This is now the negative form of the original number 00010101. *The original full number of bits must be used throughout this process.*

Example:	Binary number	– 10010110
	Subtract 1	– 10010101
	Complement	– 01101010
	Convert	– 106 (denary)
	Denary number is	– –106

Example:	Binary number	– 11011010
	Add (–1)	– <u>11111111</u>
	(Discard carry)	– 11011001
	Complement	– 00100110
	Convert	– 38 (denary)
	Denary number is	– –38

Figure 1.11 *Examples of converting signed binary numbers to denary — two alternative methods.*

Notice the important difference between signed denary numbers and signed binary numbers. The negative equivalent of +5 (denary) is simply −5 (denary); there is no change to the number, simply to the sign symbol. The binary equivalent of +5 is 00000101, and its two's complement is 11111011, which if we convert it directly to denary is 251 (denary). The denary equivalent of a negative number is always 256 − (positive number), so that the denary equivalent of the binary version of −10 is 256 − 10 = 246 (denary).

The conversion of a signed binary number to its negative denary form is done by reversing the 2's complement procedure. The binary number has 1 subtracted, and the resulting number is then complemented, as shown in Fig. 1.11.

Binary Subtraction

Binary subtraction is achieved by converting the number which is to be subtracted (called the *subtrahend*) into 2's complement form, and then *adding* this to the other number and ignoring any carry out of the m.s.b. Suppose, for example, that we subtract 12 from 15 in binary. Denary 15 is 00001111 in binary, and denary 12 is 00001100. Converting the 00001100 into 2's complement form, we get 11110100, and adding this to 00001111 and ignoring the carry from the last place gives us 00000011, which is denary 3.

On paper, this procedure looks clumsy, but it is ideally suited to

```
Number A    ... 10110110
Number B    ... 01100011
Steps:      1. Set carry bit
            2. Complement (B) ... 10011100
            3. Add to (A) + carry ...
                              10110110   (A)
                              10011100   complement (B)
                                     1   carry
          (1 carry discarded)  01010011   (result)
```

Figure 1.12 *How the microprocessor deals with 2's complement subtraction.*

microprocessor operation. Since the process which produces the final answer is an *addition* with the carry ignored, the circuits (gate logic circuits) which already exist within the microprocessor to carry out addition can be used unchanged for subtraction. This makes the design of the microprocessor considerably simpler than it would be if separate subtraction circuits had to be used. Furthermore, the complementing step is a simple command which can be carried out on a byte held in a store, and the addition of 1 to form the 2's complement can be done by setting the carry bit in the status register *before* the final addition. The procedure within the microprocessor is therefore as shown in Fig. 1.12, where the complement is found, the carry bit set, and a normal add with carry bit is executed. The carry out from this addition is then ignored.

Other Functions

In general, the only arithmetic functions which are provided in the instructions of a microprocessor are addition and subtraction. The reason is that machine control, which is the task for which most microprocessors were initially designed, does not usually require any other actions of an arithmetic type. Where these actions, of which multiplication and division are the simplest, are needed, they can be carried out in two different ways.

One is to use programs which will carry out multiplication or division by using only addition and subtraction steps. Any method of achieving an objective by a number of simple steps is called an *algorithm* (see Chapter Three) so that arithmetical algorithms are needed. Many of these have been known for centuries, and there is no need to attempt to invent new methods. Even complex operations such as finding powers of numbers (exponentiation) can be carried out by algorithms which make use of only addition and subtraction as arithmetic functions.

The other method of coping with more advanced arithmetic is to use an auxiliary chip, called a 'number-cruncher' under the control of the main microprocessor, to carry out these functions.

Summary 1.3

The negative form of a binary number is obtained by using the 'two's complement' method. The complement is the number formed by exchanging each digit for its inverse (0 for 1, and 1 for 0), and the 2's complement is obtained by adding 1 to this complement. An important consequence of this method is that the form of a negative binary number is quite different from that of its positive counterpart. Binary subtraction is achieved by taking the 2's complement of the number that is to be subtracted, and then adding it to the other binary number, ignoring any carry. The numbers must each contain the same total number of bits. Two's complement subtractions are particularly suited to microprocessor operations, because they can use the same circuits as additions.

Only addition and 2's complement subtraction are provided as commands in most microprocessors. Other mathematical functions have to be obtained by using a series of addition, subtraction and other simple operations, following the steps of an algorithm.

Exercises 1.3

1. What is meant by inverting a bit?
2. What is a complement of a number?
3. How is the 2's complement of a byte formed?
4. Form the 2's complement of (a) 00011011, (b) 01110011.
5. How is subtraction done using 2's complement?
6. Convert to binary signed form: (a) -59, (b) -22, (c) -107 (all denary).
7. Carry out in binary the subtraction 104 -32 (denary).
8. What is done with the carry bit before a 2's complement subtraction?
9. What is done with the carry bit after a 2's complement subtraction?
10. How are operations such as multiplication and division carried out in most microprocessors?

End-of-Chapter Test

1. What is meant by (a) binary number, (b) denary number?
2. Why are binary numbers used in microprocessor programming?
3. What is meant by (a) m.s.b., (b) l.s.b. of a binary number?
4. What is (a) a bit, (b) a byte?
5. What is (a) octal scale, (b) hex scale and why are they used?
6. Convert to hex and to octal: (a) 527D (b) 14208D (c) 10011011B (note D = denary, B = binary).
7. What number will be left in an 8-bit store when 11001101 is added to 11110001?
8. How is the carry bit used in a 16-bit addition when the stores of the microprocessor contain only 8 bits?
9. Where is the carry bit stored?
10. How can a negative number be represented in binary?
11. Convert to binary form: (a) −12, (b) −36, (c) −72 (all denary).
12. In what two ways could the binary number 10011011 be interpreted?
13. What is meant by (a) the complement, (b) the 2's complement of a binary number?
14. Convert and subtract, using 2's complement, 137 − 84 (both denary).
15. What is an algorithm? Give an example of where it would be needed.

2

Inside the Microprocessor

Registers

A REGISTER is an electronic circuit which can be used to store binary digits and to carry out simple operations on groups of binary digits. A microprocessor consists of several registers, which can be connected to each other or to pin contacts on the body of the microprocessor by gates which are controlled in turn by a program.

The units of a register are called flip-flops, and each flip-flop stores one binary digit (bit). An eight-bit register is one which is made up from eight flip-flops, so that it can store and work with eight bits, a single byte. This size of register is very common, but 16-bit (two-byte) registers are also used extensively in microprocessors, and some modern designs use three- and four-byte registers.

Since binary digits consist of electrical signals, they can be stored only by making use of electronic units such as flip-flops, so that registers form one very important type of storage (the other is referred to as memory). When a signal exists in the form of a voltage on a line, all that can be done with it is to gate it (switch it) or allow its voltage to affect the voltage of another point. The use of a register allows us to store a set of bits even after the voltages which they represented are no longer present.

Loading and Storing

The words 'loading' and 'storing' have special meanings when they are applied to registers. If we think for a moment of a single-byte register, then loading the register means allowing the output connections of the register to take the same voltages as exist at that moment on a set of input lines (Fig. 2.1). For example, loading a register with the byte 2CH (00101100 in binary) means that the eight flip-flops of the register have been connected to voltages which are at the levels (0 or 1) indicated by the binary form of the numbers (Fig. 2.2), so that the outputs of the register are at this same set of voltages, 00101100.

A register may be loaded with signals which have reached it along lines from various different sources. The main register of a micro-

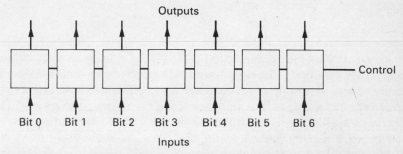

Outputs

Control

Bit 0   Bit 1   Bit 2   Bit 3   Bit 4   Bit 5   Bit 6

Inputs

Figure 2.1 *A set of units called flip-flops can be connected together to form a register. Each register will have inputs, outputs and some form of control signal to determine whether inputs or outputs are being used.*

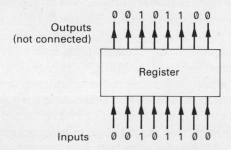

0 0 1 0 1 1 0 0

Outputs
(not connected)

Register

Inputs   0 0 1 0 1 1 0 0

Figure 2.2 *A register is* loaded *by connecting voltages to the inputs, and activating the control signal. The outputs, which are not connected during loading, take up the same voltages as the inputs, and stay at these voltages until another LOAD occurs.*

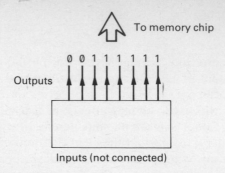

Figure 2.3 *A register is* stored *to memory by connecting its output voltages to lines which are also connected to memory inputs. The inputs of the register are disabled during this operation.*

processor (called the accumulator) is loaded by signals on a set of lines called the 'data bus'. The signals may have come from external devices (a keyboard, for example), but are much more likely to have come from electronic storage units, the memory of the microprocessor system. The action of loading a register from the memory is called 'reading'.

Storing a register means using the data-bus lines to connect the outputs of the register to other circuits, usually memory, but possibly outputs such as video display units (VDUs) or printers. For example, storing the byte 3FH from the accumulator register to memory means that the bits 00111111 are put on to the data-bus lines, making the voltages on these lines as shown in Fig. 2.3. The action of storing a byte or set of bytes from a register to memory or to external devices is also called 'writing'.

Neither loading nor storing causes any *transfer* of bits, only *sharing*, despite the terminology used. Loading a register from memory means that the register will have the same bits at its outputs as the memory has, so that the operation is one of copying. Reading is a good word to use in this respect, because the act of reading words on a page does not erase those words, any more than reading a byte from memory into a register will erase the byte from the memory. Any byte which was previously contained in the register will, however, be destroyed or 'overwritten' by this process. Similarly, when a byte is stored from a register to memory, the two will end up containing the same bits, but any byte that was previously present in the memory is overwritten.

Summary 2.1

A register is an electronic circuit which can be used for storing and manipulating groups of bits. Microprocessors consist of assemblies of registers which can be connected in various ways by gates. For microprocessors which use data numbers in single-byte form (8-bit microprocessors), both single-byte and double-byte registers will be needed.

A register is said to be loaded (in a reading operation) when the outputs of the register take the same logic voltages as a set of input lines, the data bus of the microprocessor system. Storing a register (a writing operation) means that its outputs are connected through lines to the inputs of another register or to a memory so as to affect the second register or memory. The action of loading or storing does not change the signals in the register or memory that is the *source* of the signals.

Exercises 2.1

1. What is a flip-flop?
2. What is a register?
3. How many flip-flops make up an 8-bit register?
4. What sizes of registers would an 8-bit microprocessor use?
5. Why are registers useful?
6. What is meant by loading a register?
7. What is meant by storing a register?
8. What is meant by (a) accumulator (b) data bus?
9. What is meant by (a) reading (b) writing?
10. A byte in register C is loaded into register A. What change occurs in register C?

Byte Manipulation

A byte which is stored in a register can be manipulated in that register. The types of manipulation are arithmetical or logical. Arithmetical manipulations include the addition of another byte to or the subtraction of another byte from the byte in the register. The result of either action is another byte stored in the *same* register. Taking the example of a

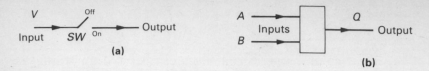

Figure 2.4 *Gating.* (a) *A simple switch gates a voltage from its input to its output when it is switched on. An electronic gate* (b) *will gate a 1 output only for a certain combination of inputs (at least two). Two inputs, marked as A and B, are illustrated here.*

A	B	Q
0	0	0
0	1	0
1	0	0
1	1	1

Figure 2.5 *A truth table for the AND gate.*

Hex		Binary
6AH		01101010
B9H		10111001
28H	result	00101000

Figure 2.6 *The AND gate operation in hex.*

single-byte register, any carry that is generated from the highest order place of an addition must be stored in another register, and similarly any carry that has to be taken to the register for a subtraction must be stored in another register. This register that is used to store the carry is called the *flag* or *status register*, and only one bit of this register is used for this purpose. Subtraction is carried out using two's complement addition; very few microprocessors have instructions for multiplication or division.

The logical manipulations include gating, shifting and rotating. Gating means comparing each bit of a byte with the corresponding bit of another byte (Fig. 2.4) and setting (to 1) or resetting (to 0) that bit of the register in accordance with the rules of the type of logical gating that is being used. The types of logic gating that are used in microprocessors are called AND, OR and XOR.

When the AND-gate logic is used, the register bit is set only if both bits were also set (to 1). The action is summarized in the *truth table* of Fig. 2.5. When this action is carried out on two bytes, the result will be a byte which has a 1 only in positions where both of the original bytes had

A	B	Q
0	0	0
0	1	1
1	0	1
1	1	1

Figure 2.7 *A truth table for the OR gate.*

Hex	Binary
A2H	10100010
5CH	01011100
FDH	11111110

Figure 2.8 *The OR gate operation in hex.*

A	B	Q
0	0	0
0	1	1
1	0	1
1	1	0

Figure 2.9 *A truth table for the XOR gate.*

1's. The name of the action comes from the description — 1 AND 1 gives 1, all other combinations give 0. The action in a complete byte is much easier to follow when each byte is written out in binary form rather than in hex — Fig. 2.6 shows the result of the AND action when the register originally contained 6AH and this byte is AND-gated with the byte B9H. The result, 28H, is *not* an answer which can be obtained by any process of arithmetic, and there is *never* any carry generated or used by this type of action. The OR-gate action is that the resulting bit is set (to 1) if either of the two original bits is set. This action is summarized in the truth table of Fig. 2.7, and shown used on two bytes in Fig. 2.8. Once again, the result is easier to follow when the bytes are written out in binary. The action gets its name from the description — the bit is set if one OR the other of the original bits was set. The OR action, like the AND action, does not generate or use a carry.

The XOR gate action is very similar to that of the OR gate, but excluding (hence the full form of the name, which is exclusive-OR) the case when both original bits are 1. The truth table is shown in Fig. 2.9,

Hex	Binary
A3H	10100011
74H	01110100
D7H	11010111

Figure 2.10   *The XOR operation in hex.*

	10110101
XOR:	10110101
	00000000

Figure 2.11   *If a byte is XOR'd with itself, the result is zero.*

its effect is that a bit is set only when one, and one alone, of the original bits is set. The effect of the XOR action on a pair of bytes is illustrated in Fig. 2.10.

These gate actions can sometimes be used in unexpected ways. For example, an XOR of the byte in a register with the same byte has the effect of clearing the register (Fig. 2.11), and an AND with FFH has the effect of leaving the register unchanged. These commands are frequently used in programs, and can be puzzling unless their action is understood.

Shift and Rotate

Unlike arithmetic or gate-logic actions, the logic actions of shifting and rotating can be carried out using only the byte that is in the register. A shift has the effect of changing the position of each bit in the byte by one place. If, for example, we take the byte 01010101 and shift each bit one place right, keeping to a total of eight bits, we obtain 00101010. A 0 has shifted into the most-significant place of the byte (the left-hand side), and the 1 which was at the least-significant end has shifted out and is no longer stored. A second right-shift will give 00010101 (Fig. 2.12), with each bit of the byte shifting along one place to the right again.

Left-shift is equally possible. If we start (Fig. 2.13) with the byte 01010101, and shift left, the result is 10101010, and a second left shift will give 01010100. Like gate actions, the results of shifts are much

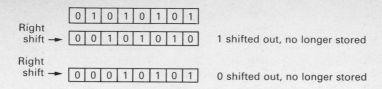

Right shift → 1 shifted out, no longer stored

Right shift → 0 shifted out, no longer stored

Figure 2.12 *Right-shift.*

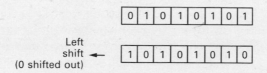

Left shift ← (0 shifted out)

Figure 2.13 *Left-shift.*

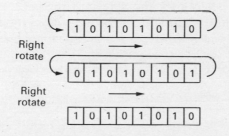

Right rotate

Right rotate

Figure 2.14 *Right rotation.*

easier to follow when the bytes are fully written out in binary. Rotation involves the same action as shifting, but with a connection made between the ends of the register, so that no bit is lost from the register. The bit which leaves one end of the register is simultaneously fed into the other end, so that all the bits of the byte are retained, though their order is changed. For example (Fig. 2.14), if the byte 10101010 is rotated right, it becomes 01010101, and a second right rotation makes it 10101010 again. A byte such as 11000011 when rotated right becomes 11100001, and a second right rotation will make it into 11110000. Like shifting, rotation can be to the right or to the left. Some types of microprocessors have shift and rotation instructions which act on 9 bits, using the carry bit as part of the register.

Instructions — Operator and Operand

The coding for an instruction to a microprocessor must normally consist of two parts. One part is called the *operator*, and is the part that defines which operation, AND, shift, etc., is to be carried out. The second part is needed by most instructions, and is called the *operand*. It consists of additional information that is needed to carry out the instruction. For example, if we use a command which adds the byte 3FH to the contents of a register, then the ADD operator must be followed by the operand, 3FH, which is the byte that is needed to complete the operation — without this byte, the instruction cannot be carried out.

The operand may, however, be a byte, or bytes, which *leads* to the correct byte being used. We can, for example, use a two-byte address as an operand, so that the command ADD,ADDRESS (operator ADD ; operand ADDRESS) will cause a byte to be read from the address given in the operand and so added to the register.

The operand may also be a code that specifies a register. For these operations which affect only the byte in a register and no other, an operator such as SLC (shift left circular) may need to take an operand which will specify which register is to have its stored byte shifted in this way.

Summary 2.2

Many register operations act on all the bits of a byte, and the two types of operations are arithmetical or logical. Arithmetical operations usually comprise addition and (2's complement) subtraction only; logic operations include gating, shifting, and rotation. Gating is an operation which compares each bit in a byte with the corresponding bit in another byte, and produces a resulting bit according to the rules of the gate action, which are AND, OR, or XOR. Shift or rotation actions do not involve any other byte, they are actions which shift the bits of a single byte either to the left or to the right. In a shift action, the bits that are shifted out of the register are lost, and zeros are shifted in. In a rotate action, a byte which is shifted out of one end of the register is shifted in to the other end of the same register.

Exercises 2.2

1. Name the two types of byte manipulation.
2. Give another name for the flag register of a microprocessor.
3. When is a carry bit used?
4. What class of operation comprises gating, shifting and rotation?
5. Write truth tables for (a) AND (b) OR (c) XOR of two bits.
6. Write results for the following (all numbers in hex) (a) B2 AND 1F (b) 07 OR 15 (c) F4 XOR 96.
7. What is meant by (a) shifting, (b) rotation?
8. What is the result (in hex) of shifting the byte 3FH left?
9. What is the result (in hex) of shifting the byte 7DH right?
10. What is meant by the operator and the operand of an instruction?

Extended Block Diagram

We can now draw a block diagram of a microprocessor which includes the features we have considered so far. Fig. 2.15 shows this block diagram, with five registers, an ALU (Arithmetic and Logic Unit), and a block which is marked Control and Timing. This block diagram is not

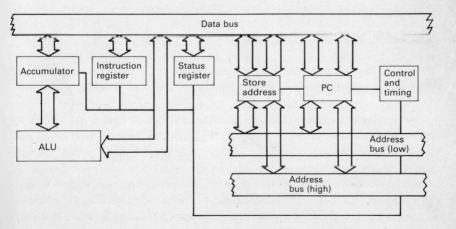

Figure 2.15 *A block diagram for a microprocessor, showing five registers, two sets of buses, and a control system.*

intended to represent the structure of any particular microprocessor, but is a simplified diagram of an imaginary one drawn to illustrate the points we have covered.

The connections between the units are of particular importance. All of the registers are able to make connections with a set of lines, called the *data bus*, which can pass signals between the units. By connecting two registers to the data bus at the same time, signals can be copied from one register to another, and this data bus, which is completely internal, inside the microprocessor chip, connects to pins and so to the data bus of the whole microprocessor system.

The accumulator is the most important of the registers, and the one which can be used the most by program commands. The accumulator can pass bytes to and receive bytes from the data bus, but it can also pass bytes to the ALU, the unit that is responsible for the actions of addition, subtraction and logic gating.

The instruction register is a more specialized register which is not directly controlled by any program written by the user — its action is entirely 'built in' by the manufacturer. Each instruction byte, the operator part of an instruction, is routed from the data bus into this register and is then decoded in the instruction register. As a result of the decoding of the operator byte (or OP-CODE) in the instruction register, the connections between the other registers are prepared so that the operand or operands can be correctly dealt with. The routing of the operator code is carried out by using the sequence of operator— operand. The first byte in any group put on to the data bus after resetting (and so restarting) the microprocessor is *always* treated as an operator and is sent to the instruction register. Thereafter, the way in which each byte is treated inside the microprocessor is decided by the type of byte that has gone before it, because when each operator is decoded, it will prepare the microprocessor to receive the correct number of operand bytes, following which the microprocessor will be set up to receive another operator.

The program counter (PC) register is an address register which must be capable of storing a full address. This will amount to 16 bits for an 8-bit microprocessor, and either 24 bits or 32 bits for a 16-bit microprocessor. The content of this register at any time during a program is the address in memory of the program step that is being carried out, and the PC will normally be incremented (its address increased by 1) after a step has been completed. After an operator has been decoded, for example, then if an operand byte is needed, the PC will increment so as to obtain the byte from the next address in the program memory. A program which is stored in the program memory can be 'run' by loading its starting address (the address number of its first instruction) into the PC.

The PC must be capable of being connected to the address bus, a set of lines which connects the microprocessor to the memory chips of the system, and it should also be possible to connect the PC to the data bus. Because the PC can always contain more bytes than can be contained on the data bus at any given time, the PC can only be loaded from the data bus in more than one step. For an 8-bit microprocessor, for example, two load operations are needed to put a full 16-bit address into the PC.

The address-store register is another full-address register which can also be connected to either set of buses. The function of the address-store register is to store addresses which are *not* in the normal sequence of the program (the addresses in the PC). In this way, a program can jump to a new address, and then return to the normal PC sequence. The address-store register can be incremented, like the PC, if required.

The status register is not a true register like the others, but a collection of individual bits which do *not* make up a byte. The bits, or *flags*, are used to indicate the status (hence the name) of other registers or memory after an operation has been carried out, and these flags can be used to control the next sequence of operations. One bit of the status register is the carry-bit which is used to store the carry from the msb of an addition. When an addition in the accumulator produces a carry out of the msb, this causes the carry bit in the status register to be set (to 1); if there is no carry from the addition, the carry bit is reset (to 0). Most microprocessor designs are arranged so that this carry bit is automatically added into the lowest place of the next addition (Fig. 2.16), so that arithmetic can be carried out on numbers which use more than one

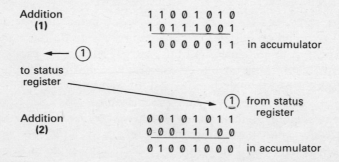

Figure 2.16 *When one addition produces a carry out of the register, this is stored in the status register and added to the lowest place in the next addition.*

byte. This automatic carry action operates correctly only if the lowest bytes to the two numbers are added first, then the next more significant bytes and so on, adding the most-significant bytes last. To comply with this, many designs are arranged so that multi-byte numbers are stored in memory with the lowest bytes at lower memory addresses, so that the normal incrementing action of the PC reads the bytes into the register in the correct (low byte first) order.

Two other bits of the status register are concerned with reporting zero and negative/positive results. The zero (Z) bit is set (to 1) if an operation in a register has produced zero; for example, when a byte is subtracted from an identical byte in a register — this is a method of recognizing a specific byte, such as the ASCII code (American Standard Code for Information Interchange) for a new-line, 0DH. For any non-zero result, the Z-bit (or Z-flag) is reset (to 0).

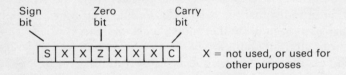

Figure 2.17 *An imaginary status register containing the three main flag bits. An overflow bit is also usually provided.*

The sign bit (sometimes called the negative bit) is set if an operation in a register produces a result which could be interpreted as a negative number — that is with its most-significant bit set. A number between 80H and FFH will therefore set this sign (or negative) bit, and a number below 80H will reset it.

These three, the C, Z and S flags, are the main status register bits which are found on all microprocessors in some form. Since the status register can accommodate eight bits, several bits can be used for other purposes, and these are used differently in different designs of microprocessors, though the full set of eight is very seldom used. Finally, all of these actions within the microprocessor need to be timed and controlled. The main timing is achieved by a 'clock generator', a circuit which generates voltage pulses that activate all sections of the microprocessor. Each clock pulse will initiate some action in the microprocessor, though several clock pulses may be needed to carry out a complete action. Microprocessor specifications will list the time for each instruction in terms of the 'clock cycle', the time from one clock pulse to another. A very common clock cycle is one microsecond, which is one millionth of a second. In addition to timing, control pulses must be sent to the various points of the microprocessor and the complete system outside the microprocessor.

An example of external control pulses is the read-write (R/W) line which is connected to the memory chips (see Chapter Four). Conventionally, this line is high (logic 1) for reading, so that when an address is put on to the address line, a byte can be read from memory. When the control unit forces this voltage low, however, the memory will be *written*. When this happens, any byte which the microprocessor has in its accumulator register will be put on to the data-bus lines, and will be written to the address in memory which has been put onto the address bus lines, perhaps from the store-address register.

The Fetch-Execute Cycle

The use of a microprocessor requires a program to be available, stored in memory chips. These memory chips are connected to the microprocessor by the data bus and the address bus, so that the microprocessor can select any address in memory and pass data along the data bus between the microprocessor and the memory chips. Any instructions, therefore, must start by fetching the instruction code byte or bytes from memory. This is called the fetch part of the cycle, and the byte which is put onto the data-bus lines from the memory is the operator byte, and is routed to the instruction register of the microprocessor.

The next part of an instruction consists of an execute cycle, when the action that has been programmed by the byte in the instruction register is carried out. If more bytes (operands) are needed to complete the action, then further fetch cycles will be needed before the final execute cycle. For example, suppose that an ADD to accumulator instruction is

Fetch instruction from memory
Decode instruction (execute)
Fetch data needed from memory
Execute instruction
Fetch next instruction ...

Figure 2.18 *The fetch-execute cycle of steps.*

to be carried out, and there is a byte already in the accumulator. The first part of the action consists of a fetch cycle, in which the ADD instruction byte is fetched into the instruction register. Once in this register, the byte is decoded to find if any further fetch cycles are needed. If we assume that the instruction was an ADD immediate, which means that the byte to be added is the next byte stored in program memory, then this byte has to be fetched before anything else can be done. A second fetch cycle is therefore started, in which the PC is incremented so that the next address in program memory is activated. The new byte is read into the accumulator at the same time as the byte already in the accumulator is transferred into the ALU, and the execute cycle continues with the addition of the new byte in the accumulator to the byte in the ALU, with the result being placed in the accumulator. When the last bit is in place, and any carry from the last place has been stored in the carry bit of the status register, the end of the instruction causes the PC to increment, so as to start the next instruction by reading in the next operator byte.

These alternating cycles of fetch and execute are repeated for each instruction. Each cycle will take a certain amount of time, measured in clock cycles, so that the time for a complete instruction may be several clock cycles, usually between four and ten on average. By knowing the time that is needed for each instruction, the total time needed to execute a program, or any part of one — a timing loop, for example — can be calculated.

Summary 2.3

A more complete block diagram of a microprocessor (compared to the block used in *Microelectronic Systems Level 1*) must include address storage and instruction decoding, and such a diagram is shown in Fig. 2.15. The accumulator is the main 'action' register, to which data bytes are loaded and in which they are manipulated. Some microprocessors have more than one accumulator.

The instruction register is used to decode the instruction bytes of the program, and so to activate the gates which connect registers together inside the microprocessor. Part of the decoding process is concerned with the correct routing of any operand of an instruction. The program counter (PC) is an address register which is used for the addresses in memory of each of the program bytes, and which in normal operation is incremented (increased by one) after each byte has been dealt with. Other addresses can be kept in one or more store-address registers, and any one of these address registers can have its outputs connected to the address bus.

The status register contains a collection (not always eight) of bits which are used to indicate the result of an operation. Typical status bits or flags are the carry bit, the zero bit and the sign (or negative) bit. The arithmetic and logic unit (ALU) is the part of the microprocessor in which all arithmetic and logic actions are carried out.

The action of the microprocessor consists of fetch and execute cycles which are controlled by the program steps. Each cycle takes a known number of clock pulses (cycles) to complete, so that its total time can be calculated precisely.

Exercises 2.3

1. Mark on a blank block diagram (a) the ALU (b) the PC.
2. What is the purpose of the instruction register?
3. How is the program counter (PC) used?
4. Why is a store-address register needed?
5. How is a store-address register used?
6. What is the purpose of the status register?
7. Name three bits (flags) which would be found in a status register of any microprocessor.
8. What is the clock generator?
9. What is meant by the fetch/execute cycle of the microprocessor?
10. How can the time for completing a program loop be calculated?

End-of-Chapter Test

1. What is meant by (a) flip-flop, (b) register?
2. What are the uses of registers?
3. What is meant by (a) loading (b) storing a register?
4. What is the name of the set of lines that can be used to link the accumulator with the memory?
5. A byte is loaded from memory into the accumulator. Is there a change in the memory or in the accumulator?

6. Give an example of a logical manipulation.
7. Give the (hex) results of the following operations: (a) 0BH AND F8H (b) 56H OR 3DH (c) F2H XOR A5H. *a ~08 b = 7F. C =57*
8. Give the (hex) results of (a) left shift (b) right shift of 6AH.
9. Give the (hex) results of (a) left rotation (b) right rotation of 3BH.
10. State what is meant by (a) operator (b) operand of an instruction.
11. What is the function of the ALU in a microprocessor?
12. Why is the accumulator regarded as the most important register?
13. How is the program counter used in a microprocessor?
14. How are the bits in the status register used?
15. Give a simple example of fetch and execute cycles of an instruction.

3
Creating Programs

Algorithms

How do you solve a problem? Whatever the problem is, simple or difficult, your method of solution must make use of alternate steps of getting information and using it. When we write these steps down, then if each step is a simple one, what we have is an *algorithm* for that problem — a method of solution using only simple steps. Sometimes, of course, our attempts to find an algorithm may indicate that no solution can actually be found.

Consider an example that at first sight looks simple — arranging a set of ten numbers in order from lowest to highest. You can do this easily just by inspecting the numbers, and writing an order 1, 2, 3 and so on, against each one, but how is this done step by step, in the way a machine would do it? Fig. 3.1 shows one of several possible methods. The first number in the list is put into a store, and eliminated from the list. The next number is then compared with the one in the store (the first number). If the second number is less than the one in the store, it is exchanged with that number, so that the smaller number of the two is now in the store, and the greater is back in the list. If, on the other hand, the second number is larger than the one in the store, no change is made. The comparison is now made again, comparing the third number in the list with the number in the store, and exchanging only if the stored number is larger than the number in the list. When all the numbers in the list have been dealt with in this way, the number in the store must be the smallest number, since it has been compared with all the others, and only the smaller number in any comparison is left in the store. At the

1. Take first number from list and put in store.
2. Take next number from list and compare with stored number.
3. If list number is less than the stored number, exchange numbers, else leave.
4. Repeat with new store position until all numbers from list are in store.

Figure 3.1 *An algorithm for sorting numbers.*

end of this set of operations, the list is one less than it was, and the smallest number is in store. Now if we repeat this set of actions, using the remaining list, and using for our store the *next* store address, we end up with two numbers in store which are in ascending order, and if we keep going until all the numbers in the list have been transferred to the store, then the numbers in the store will be in ascending order (Fig. 3.2).

This is slow, tedious and repetitive work for a human to do — but that is just the kind of work that is suited to a microprocessor system. A microprocessor can carry out these simple compare-and-exchange steps very rapidly, even faster than the human brain can sort ten numbers. Unless we can write an algorithm, however, we cannot program the microprocessor to carry out the task. An algorithm is sometimes called a method which can be used by a fast idiot — and 'fast idiot' is not a bad description of a microprocessor.

Algorithm to Flowchart

It is not easy to write a program directly from a written algorithm, though some microprocessor programmers have a talent for doing this. Most of us need the algorithm to be put into a clearer form, and for many programmers, a flowchart is a satisfactory form for this purpose.

A flowchart consists of a set of symbols which show the steps of an algorithm and how these steps are repeated in a program. You cannot make much use of a flowchart until you know the standard flowchart symbols, so let us start by explaining each symbol and its use, referring to the symbol chart of Table 3.1. The Start and Finish symbols need little explanation. Each algorithm must have some starting point at which its action starts, and some finishing point at which all the action is complete. Marking these points makes it easier to follow the progress of the algorithm in its flowchart form. The Input/Output symbol indicates where data are read into or written out from the microprocessor system.

	LIST	STORE	Comments
(a)	16		Start—list full, nothing
	2		in STORE
	7		
	1		
(b)	LIST	STORE	
	2	16	First number from LIST
	7		transferred to STORE
	1		
(c)	LIST	STORE	
	16	2	2 less than 16, so
	7		exchange
	1		
(d)	LIST	STORE	
	7	2	7 is greater than 2, so
	1	16	no exchange
(e)	LIST	STORE	
	16	2	1 is less than 2, so
	7	16	exchange this time
	2		
(f)	LIST	STORE	
	7	1	Take next number
	2	16	to STORE
(g)	LIST	STORE	
	16	1	7 is less than 16, so
	2	7	exchange this time
(h)	LIST	STORE	
	16	1	2 is less than 7, so
	7	2	exchange this time
(i)	LIST	STORE	
	7	1	Take next number to STORE
		2	
		16	
(j)	LIST	STORE	
	16	1	7 is less than 16, so
		2	exchange this time
		7	
(k)	LIST	STORE	
		1	Last number taken over,
		2	end of sort because LIST
		7	is now empty
		16	

Figure 3.2 *How the sort algorithm operates for four numbers.*

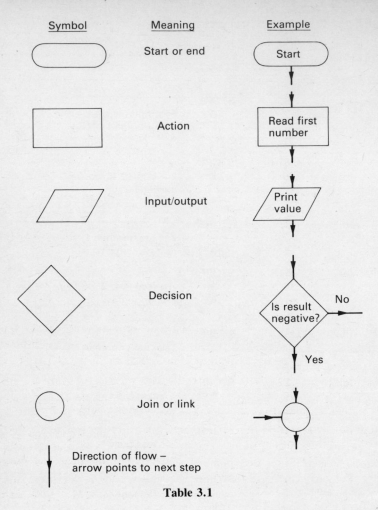

Symbol	Meaning	Example

Table 3.1

This does *not* apply to reading or writing memory. What it refers to is an input from, for example, a keyboard, at some part of a process, or an output to, taking another example, a VDU screen or a printer at some other stage in the program. Many programs make no use of Input/Output stages.

The Process symbol indicates the main 'doing' steps of the algorithm. Steps such as comparison of bytes, read from memory, write to memory, addition (and all other arithmetic), logic gating, shifting and rotating or testing the values of bits are all actions which would be indicated by this symbol, which makes it the symbol that is used most often in the course of a flowchart.

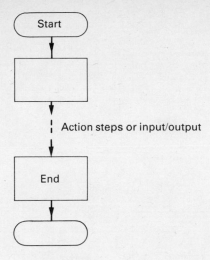

Action steps or input/output

Figure 3.3   *A linear flowchart — there are no decision steps.*

The other remaining symbol, the lozenge shape, indicates a decision step from which the program flow can proceed in one of two separate ways. The usual way of expressing the decision is as a question written inside the symbol, with the lines coming from the symbol marked as Yes and No according to the answer to the question. This implies that the question in a decision step must be one that can be answered by using Yes or No only, because these are the only forms of decisions that the microprocessor can deal with (two choices, one of which can be indicated by 0, the other by 1).

A flowchart can be linear or branching. A linear flowchart (Fig. 3.3) consists of a set of steps from Start to Finish, with no decision steps, so that there is only one possible path from Start to Finish. A branching flowchart (Fig. 3.4) has decision steps (branches) so that there may be several possible paths between the Start and the Finish (two at least if there is one decision step). One very useful type of branching flowchart is the looping or iterative flowchart in which a set of steps is repeated until the result of a decision step changes, causing a different path to be followed (Fig. 3.5).

The flowchart is not the only way of arranging the steps of an algorithm so that they can be transformed into the steps of a program,

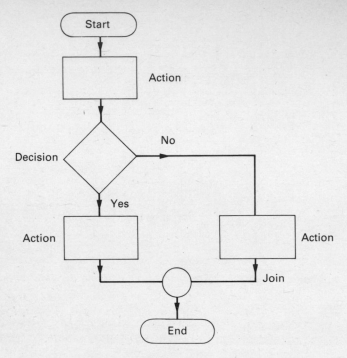

Figure 3.4 *A branching flowchart.*

but it is a relatively simple method which is easy to learn and use. Another advantage of the flowchart is that its shapes are easily recognized so that a faulty step can often be seen by a brief inspection.

Summary 3.1

Because each action of a microprocessor is a very simple one, any problem that is to be solved by using a microprocessor must be broken down into a number of simple steps. A method of approaching a problem in simple, usually repetitive, steps like this is called an algorithm. Once an algorithm for a problem has been worked out, a flowchart can be drawn up, unless the algorithm has shown that the problem cannot be solved. A flowchart is a pictorial form of the algorithm which shows at what parts of the program data must be put in or taken out, where decision steps exist and where action steps are placed. The flowchart is drawn using standard symbols.

Flowcharts may be linear or branching. A linear flowchart consists of

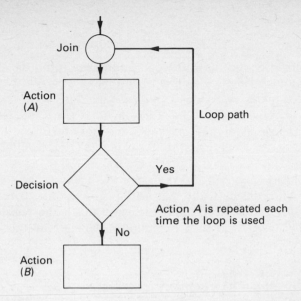

Figure 3.5 *The appearance of part of a looping flowchart. The branch is to an earlier part of the program, so that the actions within the loop (action A and the decision step in this example) are repeated until the other path (the NO decision in this example) is taken.*

a single set of steps from start to finish, a branching flowchart has decision stages which permit several different possible paths to be taken between the start and the finish. A looping flowchart is a type of branching flowchart in which a part of the flowchart is used several times over in the course of completing the action.

Exercises 3.1

1. What is an algorithm?
2. Why is an algorithm needed?
3. Is the procedure for converting a denary number to binary an example of an algorithm?
4. What is a flowchart?
5. Why do we need a flowchart?
6. Draw the flowchart symbols for (a) Input/Output (b) Process (c) Decision.
7. Give an example of a 'process' step as applied to a microprocessor.
8. What are the possible answers to a decision step?
9. Name the two main patterns of flowchart.
10. What is a looping flowchart?

Deriving a Flowchart

As an example of the derivation of a flowchart from an algorithm, consider the very simple example algorithm of Fig. 3.6, which is an algorithm for taking two bytes from separate store addresses, adding them, then putting the result back into another store address. Rather than using imaginary address *numbers*, the store addresses have been represented by the labels or symbolic names STOR1, STOR2, STOR3 respectively.

The flowchart (Fig. 3.7) shows in more visual form the steps of the algorithm. After the start, the first action is to read the byte stored in address STOR1 into the accumulator, since this is the register that is most usually involved in such operations. Having read this byte, the

1. Take byte from STOR1.
2. Take byte from STOR2, add to previous byte.
3. Store result in STOR3.

Figure 3.6 *An algorithm for the addition of single bytes, single byte result.*

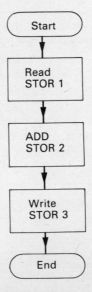

Figure 3.7 *A flowchart for the simple addition — this is a linear flowchart with only three action steps.*

next step is to add the byte in STOR2; the ADD command implies that the byte will have to be read from memory and added to the byte in the accumulator. The result of the addition will appear in the accumulator, and the third step of the algorithm is to write this result (a single byte in this example) into address STOR3. This ends the flowchart, which is a linear one.

Each item of the flowchart can now be converted into an instruction or set of instructions to a microprocessor. Precisely what these instructions will be depends on the type of microprocessor that is used, but the same flowchart applies, because the actions must be the same. One of the benefits of drawing flowcharts is that the flowchart can be used to draw up programs for entirely different microprocessor types, whereas it would be extremely difficult to take a program written for one type of microprocessor and convert it to a form that could be used on another type.

We assumed in the example of Fig. 3.6 that the two numbers that were added were single-byte numbers and that the result was also a single-byte number. Suppose, however, that the two numbers are single-byte numbers, but that they produce a carry? Fig. 3.8 shows the amended algorithm which is now needed. We have to check now whether there is a carry from the addition, and we have to put this carry bit into another location in memory, STOR4. Whether there is a carry or not, the single-byte part of the result (the least-significant byte) is put into the same STOR3 location.

Fig. 3.9 shows the new flowchart drawn for this algorithm. The first two steps are the same as for the previous flowchart, but immediately after the addition has been carried out, there is a decision step — is there a carry? If there is a carry, then the path that is taken puts an extra step into the program — loading a bit into the address STOR4. Whether there is a carry or not, the next step is the same, loading the single-byte result from the accumulator into STOR3.

Note that if the bytes are stored in order of their STOR numbers (so that STOR1 represents the lowest address, STOR4 the highest), then

1. Take byte from STOR1.
2. Take byte from STOR2, add to previous byte.
3. Store result byte in STOR3.
4. Store carry (if 1) in STOR4.

Figure 3.8 *An algorithm for an addition with a possible carry.*

the answer, which is in STOR3 and STOR4 is stored with its lower byte in the lower-numbered memory, and its higher byte in the higher-numbered memory. Most microprocessors store numbers in memory in this order.

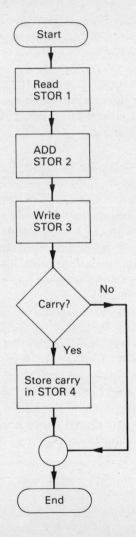

Figure 3.9 *A flowchart for the 'add with possible carry'. Note that in this flowchart, the carry is stored only if it is 1. This is not strictly necessary, since simply storing the carry bit (which is 0 if there is no carry) would suffice, but this flowchart has been used to illustrate the use of a decision step.*

An Example of Two-Byte Addition

Fig. 3.10 shows an algorithm for carrying out a two-byte addition, using only single-byte registers. Several microprocessors permit two-byte addition directly, making use of 16-bit registers, but the problem of what to do if numbers consist of more bytes than can be contained in one register is applicable to any type of microprocessor. The procedure of the algorithm is to read in the low byte of one number, then add to it the low byte of the second number. If this causes a carry, the carry bit is temporarily stored (where?) and the one-byte result, which is the low byte of the answer, is stored. The high bytes are now treated in the same way, but the carry bit is added in. The result of adding the high bytes is also put into memory and any carry from this addition is also stored in memory.

The algorithm gives rise to a much larger flowchart (Fig. 3.11) and, in addition, we need a chart of the memory locations to keep track of the way in which we are using them. The flowchart starts, as before, by loading a byte from STOR1, the low byte of the first number. The next step is to add the low byte of the second number from STOR3, to give an answer byte in the accumulator and possibly a carry bit. If there is a carry bit, it will be stored in the status register, and the program must arrange for the answer byte in the accumulator to be stored in STOR5. This is the lowest byte of the answer.

The high byte of the first number is now read in from STOR2, and the high byte of the second number added, along with any carry from the first addition. The byte resulting from this addition is stored in STOR6, and any carry from this addition is stored in STOR7, completing the action.

This flowchart has shown two decision steps relating to the carry, but in practice microprocessors are constructed so that the storage and

1. Take low byte of first number.
2. Add low byte of second number.
3. Store result byte, which is low byte of answer, and carry.
4. Take high byte of first number.
5. Add high byte of second number, plus carry.
6. Store result byte, which is next byte of answer.
7. Store carry, which is most significant bit of answer.

Figure 3.10 *An algorithm for a two-byte addition.*

addition of a carry from one ADD to the next will be done completely automatically, so that there is no need to arrange for this action in the program. Once we know that this is so, we could omit it from flowcharts, but we must remember that the carry will be added unless for some reason (such as having two quite separate additions) we do not want to use it. It is safer to put the action into the flowchart as a reminder. Note that the final writing of the carry bit into memory is *not* an automatic action, it must be programmed.

Summary 3.2

When an algorithm is being worked out, it is useful to use labels (symbols), which are short names, in place of addresses or data bytes, since this makes the flow of the action easier to follow and understand. When the algorithm has been checked (is each step a simple one?), the flowchart can be drawn. This flowchart should not be too specific — it should permit a program to be written for any type of microprocessor.

Some of the actions which are put into the algorithm and into the flowchart can be carried out by the microprocessor without any intervention by the programmer. An example is the addition of a carry from one addition step to the next. These actions ought nevertheless to be shown in the algorithm and in the flowchart as a reminder that they are being carried out.

Exercises 3.2

1. Why are label names used in algorithms and flowcharts?
2. What quantities may be represented by label names?
3. Write an algorithm for subtracting one single-byte number from another, assuming that the result is positive.
4. Derive a flowchart from your answer to Question 3.
5. Why is a flowchart at this stage more useful than a program obtained directly from an algorithm?
6. What is the advantage of storing a two-byte number with its lower byte in the lowest address number?
7. Devise an algorithm for taking a number from memory and multiplying it by two (look at the effect of a left-shift).
8. Draw a flowchart for the algorithm of Question 7.
9. Devise an algorithm for taking a number from memory, multiplying it by two, then adding a second number and storing the result. Assume single-byte numbers and answer.
10. Draw a flowchart for your algorithm in Question 9.

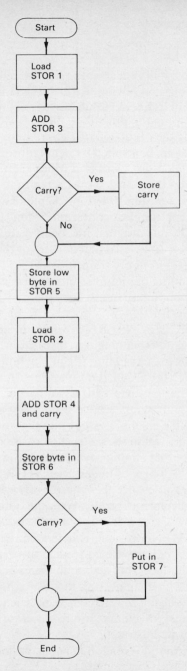

Figure 3.11 *A flowchart for adding two two-byte numbers.*

Tracing the Action

Even the execution of a program as simple as that of Fig. 3.7 involves a large number of steps within the microprocessor. The action of the microprocessor will be much better understood if each set of these steps is traced out, showing the quantities that are stored in each register at every stage during the execution of the program. These individual steps of each instruction are controlled by a 'microprogram' which is built into the design of the microprocessor and which cannot be altered in any way by the programmer. Fig. 3.12 shows a typical *trace table*, which shows the contents of the registers noted in Chapter Two. The table starts with all registers except the PC containing 'don't care' values, because nothing relevant to our program has been read into them. The address in the PC, however, is the assumed address of the first instruction of the program as it is stored in program memory.

The effect of putting this address into the PC is to transfer the address on to the address-bus lines, and to activate the READ control line. As a result, the first instruction will be read from memory, and we have assumed that this is a byte BAH which in this (imaginary) micro-processor is the read-from-memory instruction. In step 2 of the trace table, this byte has to be read into the instruction register to be decoded, as a result of which the PC increments (step 3), and another fetch cycle starts, bringing the low byte of the address of STOR1 from the next address in program memory where it is stored as part of the program. This byte, 14H in the example, is put into the *store-address* register, and the instruction code, still in the instruction register, then causes the PC to increment again so that another byte is read from program memory, this time to form the high byte of the address STOR1. Note that it is the job of the programmer to ensure that these bytes are placed in the program memory and in the correct order. With the address complete (step 4), the store-address register is connected to the address lines in place of the PC, and the READ control line is activated. In this way, the low data byte 06H from STOR1 (whose address in the example is 1214H) is read in, and is loaded into the accumulator. Following this, control is returned to the PC by reconnecting the PC to the address lines. In step 6 the PC increments again, to fetch a new instruction byte, DA, the ADD instruction of this imaginary microprocessor. This, like the load instruction that was illustrated, has to be followed by two bytes read from program memory to form an address, so that steps 7 and 8 result in the address for STOR2, assumed to be 2215H, being loaded into the store-address register. This address is then put on to the address lines, fetching the byte in 2215H (the byte 12H) in step 9, and starting the addition steps of transferring the byte from the accumulator to the ALU, putting the new byte in the accumulator, adding and placing the result, bit by bit, into the accumulator, so that by the end of step 10, the result 18H is in the accumulator register.

	R	PC	SA	A	ALU	S	Comments
1.	xx	7F00	xxxx	xx	xx	xx	Starting address in PC
2.	BA	7F00	xxxx	xx	xx	xx	BA is first instruction byte to be decoded.
3.	BA	7F01	xx14	xx	xx	xx	Fetch low byte of address.
4.	BA	7F02	1214	xx	xx	xx	Fetch high byte of address.
5.	BA	7F02	1214	06	xx	xx	Byte from STOR1 address.
6.	DA	7F03	1214	06	xx	xx	Next instruction byte.
7.	DA	7F04	1215	06	xx	xx	Low byte of STOR2 fetched.
8.	DA	7F05	2215	06	xx	xx	High byte fetched.
9.	DA	7F05	2215	12	06	xx	Byte fetched, first byte transferred to ALU.
10.	DA	7F05	2215	18	xx	xx	Addition done.
11.	FA	7F06	2215	18	xx	xx	Next instruction.
12.	FA	7F07	2216	18	xx	xx	Low byte next STOR.
13.	FA	7F08	2316	18	xx	xx	High byte STOR3.
14.	FA	7F08	2316	18	xx	xx	Answer byte stored.
15.	00	7F09	xxxx	18	xx	xx	End command.

Note: x means value unimportant.

R – Instruction register	PC – Program counter
A – Accumulator register	SA – Store address
S – Status register	ALU – Arithmetic and logic unit

If there had been a carry at step 10, the content of the status register would have changed.

The addresses which have been assumed are:

STOR1 1214	STOR2 2215	STOR3 2316

Figure 3.12 *A trace table for a single-byte addition. This is a much more detailed way of documenting program steps, which is used only when there is some doubt about the effect of an action.*

At step 11, with the address bus connected to the PC again, the PC increments to find another instruction, this time the STORE instruction. This, as before, is read into the instruction register and causes two more fetch steps, getting in from the program memory the bytes of the STOR3 address, 2316H. This address is then put on to the address bus lines in the usual way, and because a *store* instruction is in the instruction register, the memory-write control signal is sent out and the data byte 18H (the result of the addition) is written to address 2316H from the accumulator register. This concludes the program action, and step 15 is an assumed return to a waiting state (a monitor program).

A trace table for even such a simple set of instructions is very lengthy, but once the trace for the basic steps of fetch and execute is understood, there is seldom any need to examine a section of program in such detail.

The ROM

Looking carefully at the trace table emphasizes one very important point. A microprocessor is incapable of any action until some program exists to issue the instructions in the correct sequence. The trace table shows some of the action that is concerned with an addition program, but it tells us nothing of what program effort that was needed to ensure that the bytes of the program, which in the example, started in program memory at 7F00H, were put into memory.

It is, in fact, normally impossible to place a program into the memory of a microprocessor system unless there is already a program present! If this seems like a paradox, remember that a program is needed to read bytes from a keyboard, and similarly a program is needed to read bytes from memory and display them on seven-segment displays or on a VDU cathode-ray tube. No matter how simple the microprocessor system, some sort of program must be available in the first addresses of memory that the microprocessor will use when it is switched on.

This program, often referred to as a monitor or 'bug', can be fairly elaborate, containing routines for the use of a keyboard, VDU, printer, cassette recorder and other peripherals; it may even contain programs that allow commands to be typed on a keyboard using a 'high-level language', similar to English. On the other hand, it may be only enough to contain commands that permit a larger program to be read in from disc or tape, in which case it is often called a 'bootstrap' monitor.

The memory which contains the monitor program has to be permanent, not one which, like the RAM memory, will lose its stored data when the unit is switched off. The type of memory which is used for this purpose is ROM — Read Only Memory — which is made in the form of an IC chip (or several chips), with the program 'built in' during the manufacturing process.

In this way, the programs that are needed to make the microprocessor system work can be incorporated into a part of memory which is always connected, and whose starting address will be the first one that the microprocessor goes to when it is switched on. A working system will need some RAM memory in addition, however, because in the course of any program, as we have seen, bytes will have to be placed in memory for temporary storage.

A machine-control microprocessor system would probably make use of a large amount of ROM and only a small amount of RAM for temporary storage. A system which is to be used for writing new programs, on the other hand, needs only enough ROM to service the use of the keyboard, display and record/replay routines, but enough RAM to allow space for programs and for the temporary storage that these programs will need. Finally, Fig. 3.13 shows a block diagram of a system using ROM, RAM and port in/out, a complete small microprocessor system in fact.

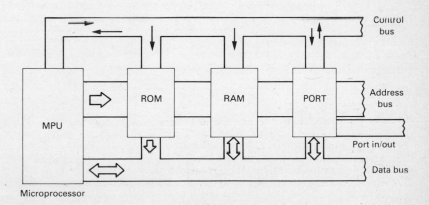

Figure 3.13 *A block diagram for a system with a microprocessor controlling ROM, RAM and a port chip.*

Summary 3.3

Because even a simple program step involves a large number of actions within the microprocessor, we need a trace table to analyse what happens within the microprocessor. This set of actions is carried out under the instructions of a 'microprogram' which is built into the microprocessor and which cannot be altered by the user. The microprogram consists of *fetch* and *execute* steps, in which data are either moved from place to place (fetch) or operated on (execute). Data are fetched from memory by putting an address on to a set of address lines. This can be done by connecting the lines of the address bus to the outputs from the program counter register or the outputs of the store-address register. The fetched data can be routed either to the instruction register for decoding, or to the accumulator (or other execution register) for carrying out an action. The correct sequence of instruction and data must be rigorously maintained, with the first byte of a set being the instruction byte.

A program is usually contained in memory, from which it can be read as fast as the microprocessor can use it. The type of memory called ROM (Read Only Memory) is important for this purpose, because its contents cannot be altered by the microprocessor action or by switching off power. RAM memory which can be altered is also needed.

Exercises 3.3

1. What is a trace table?
2. Why do we not draw a trace table for every instruction we use?
3. For what purpose is the PC used?
4. What is done in the instruction register?
5. How is the store-address register used?
6. How is an instruction byte distinguished from a data byte?
7. Why must there always be a program in memory when a microprocessor is activated?
8. What is a 'monitor' or 'bug', program?
9. What is ROM?
10. What is RAM?

End-of-Chapter Test

1. What is the first stage in solving a problem by the use of a microprocessor system?
2. Why must the problem be broken down into simple steps?
3. What is a flowchart?
4. What is the advantage of drawing a flowchart?
5. What is meant by (a) linear (b) branching (c) looping flowcharts?
6. What answers are acceptable in a decision stage?
7. How are address and data byte numbers represented in flowcharts?
8. Does one flowchart symbol correspond to one microprocessor instruction?
9. Draw an algorithm and flowchart for adding two single-byte numbers, assuming that the answer is also single-byte. Assume that the original numbers and the answer will be stored in memory.
10. What extra step(s) do we need when the answer in Question 9 is of more than one byte?
11. What is the purpose of a trace table?
12. What is meant by (a) fetch (b) execute cycles?
13. What is meant by (a) ROM (b) RAM?
14. Why is ROM essential to a microprocessor system?
15. What is a monitor program?

4

Memory Systems

Memory

As has been noted in Chapter Three, memory can be of the ROM or RAM variety. The main difference is that the RAM is read/write memory, which can be both read by the microprocessor (byte copied from memory to a register of the microprocessor) or written by the microprocessor (byte copied from a register of the microprocessor into memory). ROM, as was explained in Chapter Three, is memory which can only be read, not written by any microprocessor action, and which is unchanged when the power is switched off.

A complete memory system for a microprocessor, whether it be ROM or RAM, must usually consist of several storage devices. RAM can be of two types, static RAM, made from the electrical switches called flip-flops (like microprocessor registers), and dynamic RAM, which uses the principle of charge stored in a capacitor. Whichever type of storage device is used, each unit can store one bit, 0 or 1. A complete memory store must be able to use groups of stored bits, however; groups of eight for an 8-bit microprocessor, and groups of sixteen for a 16-bit microprocessor. Each group of bits must be located so that all the bits of that group can be accessed at the same time by using an address number.

Suppose, for example, that we have 32 storage units (using denary numbers) which can be made into four sets of 8 bits. Rather than have 32 leads coming from this (very small) memory, we arrange the storage units so that their outputs can be connected to eight lines, the data-bus

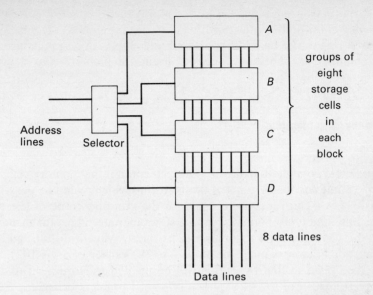

Figure 4.1 *How addressing is used to select units of memory.*

lines, so that all four sets of units will use the same eight output lines. In addition, we do not need to use four lines to switch the four sets of units, because two lines along with a binary number decoder will give us the four outputs we need for inputs of 00, 01, 10, and 11 in binary. The total is therefore ten lines, eight of data and two of address. A block diagram of this memory looks as in Fig. 4.1, with the eight data lines and two address lines that are needed to locate each set of bits.

What happens? With 00 on the address lines, line *A* of outputs from the storage units is connected to the data lines, and if the microprocessor is connected also to these lines, then this memory address can be read, with the voltages on the data lines affecting the inputs of one of the registers of the microprocessor. If we change the address number to 01, then set *B* is connected to the data lines and its different set of bits will be put into the microprocessor register. The combination of two address lines and eight data lines allows us to deliver a set of eight bits from any one of four different addresses *at random*.

Suppose that we had 2048 storage units (or cells) arranged into 256 groups of eight. We could once again use eight data lines, but to select 256 different sets does not need 256 lines, as you have probably gathered, but just eight, because 2 to the power of 8 is 256, and a binary decoder within the memory chip will be able to switch 256 different sets of storage cells using input signals from only eight lines. By putting the decoding of the address lines within the storage (memory) chip itself, we avoid having to use an additional decoding chip with an unmanageably large number of connections between it and the memory chip it serves.

Memory Organization

Though it is possible to make chips which contain the memory cells and the decoding for 1024 groups of 8 bits (a chip which would be described as a 1024 × 8 chip) it is more usual for the grouping to be of less than eight bits. One very popular type of static memory chip, for example, uses a 1024 × 4 arrangement. This unit will have ten address connections (address pins), because 2 to the power of 10 is 1024, and four data pins, so that two chips will be needed to store a set of 1024 complete bytes.

The arrangement of two such chips to supply a 1024 × 8 bit memory is shown in Fig. 4.2. The corresponding address pins of the chips are connected together, so that any address that is placed on the address lines will activate four memory cells in each chip. The four data pins of each chip are connected so as to make up an 8-line data bus. For any address placed on the ten address lines, one complete byte of data will be switched on to the data lines, four bits (sometimes called a 'nibble' — because it is half of a byte) from each chip. This organization of a given

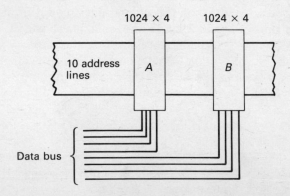

Figure 4.2 *Arranging two 1024 × 4 memory chips to give 1024 × 8 memory units.*

amount of memory, in this example 1024 bytes (called 1 kilobyte, written as 1 K) is just as acceptable as the use of a 1024 × 8 chip, and easier from the point of view of manufacturing the chip.

Another type of chip which is very widely used is a dynamic RAM chip which is arranged as 16K × 1, meaning that there are 16 × 1024 = 16, 384 single-bit storage units in each chip, along with the decoding circuits. A chip with 16K addresses will need 14 address lines, but because only a single bit is stored at each address in one chip, there will be only one data pin on each chip. If we want to use these chips to provide a complete byte of storage, then eight chips must be used, with corresponding address pins connected in parallel, and with each data pin connected to a different data-bus line (Fig. 4.3). At the time of writing, these chips provided the best value for money in terms of memory size, providing that the microprocessor system could supply the necessary signals for dynamic memory. Dynamic-memory chips, being based on the principle of charged capacitors, lose information because of leakage of charge from the capacitors, so that the charged capacitors have to be recharged or *refreshed* at frequent intervals, about every millisecond. This refreshing process can be carried out by circuits operating independently of the microprocessor, but at least one type of micro-processor (the Z-80) contains its own refresher circuits within the chip. The advantage of using dynamic memory, apart from low cost, is that the power consumption of a dynamic memory chip is much lower than that of a static memory chip with the same memory capacity.

16K × 1 bit does not represent by any means the maximum packing of memory cells that can be achieved, and at the time of writing 64K × 1 bit chips are available, though at a price which does not make them competitive. By the time you read this, however, this may no longer be true!

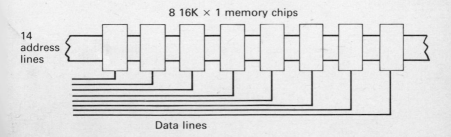

Figure 4.3 *Arranging 8 16K × 1 memory chips to give a 16K × 8 memory.*

The rules of memory organization are:

1. Each address number should be unique, so that a set of bit can be gated onto the data lines by one address and one only.
2. Each address number will gate the same number of bits (usually 8 or 16) on to the data lines.
3. There is never more than one set of bits connected to the data bus at a given time.

Provided that these rules are obeyed, any form of memory organization is acceptable.

Summary 4.1

The memory of a microprocessor system will consist of both ROM and RAM. RAM can be static or dynamic, using different principles for storage. Memory chips contain decoding circuitry which allows the memory cells (units) to be activated by a binary number placed on the address pins of the chip. Memory chips are not necessarily available with their cells arranged in groups of eight to form one byte. A single cell at each address number, or a group of four (a nibble) are more common, and the circuits have to be arranged to make use of whatever arrangements of memory is used.

Any type of memory must be arranged so that each address number placed on the address lines will activate one complete byte of memory cells and one only.

Exercises 4.1

1. What two types of RAM can be obtained?
2. What purpose do the address pins of a memory chip serve?
3. A memory chip need not have 8 data pins. Why not?
4. Why must a memory chip contain binary decoding circuits?
5. How would a 1024 × 8 chip differ from a 2048 × 4?
6. How many address lines would a 4096 × 1 chip need?
7. How many 16K × 1 chips are needed to make a 16K × 1 byte memory?
8. What is meant by a 'nibble'?
9. What is meant by 'refreshing'?
10. State the rules of memory organization.

Memory Selection: Address Decoding

An eight-bit microprocessor (with 8 data lines) normally uses 16 address lines, so that 2-to-the power 16 (65 536 bytes, or 64K bytes) can be used directly. Sixteen-bit microprocessors, with 16 data lines, use either 24 or 32 address lines, and are therefore capable of addressing much greater quantities of memory (can you work out how much?). In either case, we often need to be able to use more memory than can be supplied by one set of chips. This is done by techniques called address decoding.

For example, suppose we have 1K of memory, made up from a single 1K × 8 chip. How can we make use of two such chips to give a 2K × 8 memory? Obviously, the data pins of each chip will connect to the corresponding data lines of the data bus, but how are the address lines to be connected?

If we simply connect corresponding address pins together, then two sets of data will be connected to the data bus as each address number, and that is not permissible. Each chip needs ten address lines, and the microprocessor (assumed 8-bit) has a total of 16, so that we cannot simply use a different set of address lines for each chip. To understand how this problem is solved, we need to look closely at the sequence of signals on the address lines.

Take a simple example first, with only four lines and two chips which need only three lines (Fig. 4.4). Suppose chip *A* is connected to the lowest three address lines A0, A1 and A2. The sequence of signals, for consecutive addresses on the four lines A0, A1, A2, A3 is shown in Fig. 4.4, from which you can see that eight separate addresses will be sent to chip *A*, and then these addresses will be repeated in exactly the same sequence, but with the unused line A3 set high for another count of eight. With only one chip connected as shown, then the address numbers simply cycle round endlessly.

The separation between the two chips that we want occurs when line A3 goes high, and this is the clue to the type of circuit we need. If we connect both chips to address lines A0, A1, A2, and we use line A3 to switch the chips, so that chip *A* is active only when line A3 is low and chip B is active only when line A3 is high, the chip A will be active for addresses 0000 to 0111, and chip *B* will be active for addresses 1000 to 1111, using binary numbers.

Systems of this type are called address decoding systems, and are necessary when a large amount of memory has to be provided, using a large number of chips. To extend a 1K × 8 memory to 2K × 8, then, we connect a second 1K × 8 chip in parallel with the first one as far as data

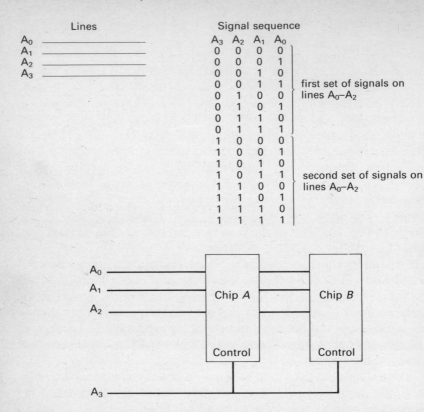

Figure 4.4 *Using an address line to select between two memory chips which are connected to the same set of lower address lines.*

and address lines are concerned, and we gate the chips with the next unused address line. 1K chips need 10 memory lines, which are numbered from A0 to A9, so that the next unused memory line is A10, and this is the one we need to use for gating. With line A10 low, one memory chip is active, with line A10 high, the other memory chip is active.

Chip Select/Enable

If address decoding is to be usable, we need some way of activating each memory chip when it is needed. This is done by means of a chip-select (or chip enable) pin on each memory chip, arranged so that at one logic

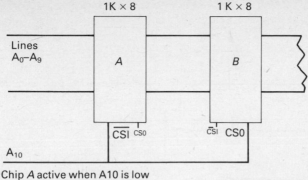

Figure 4.5 *Simple address decoding for two chips only.*

voltage the chip is completely isolated, unaffected by either address or data signals. Very often, manufacturers of memory chips will arrange more than one chip select/enable pin, so that external gating is reduced to a minimum. One commonly used procedure is to have two select pins labelled as CS0 and $\overline{CS1}$. The bar over the CS1 implies that the chip is selected when this pin is at logic 0, whereas the absence of a bar over the CS0 means that the chip is selected when this pin is at logic 1. Both pins must be at their correct logic levels to select the chip, so that the chip is active only when CS0 = 1 and $\overline{CS1}$ = 0. The nomenclature CE and $\overline{CE}$ (chip enable) may also be found.

The provision of several chip-select/enable pins is useful because it allows more complete address decoding than our simple scheme of Fig. 4.5. In the drawing of Fig. 4.5, line A10 was used to select between two memory chips, but no other line was used. Because of this, the decoding is only *partial*, and the sequence of addresses that can appear on lines A0 to A10 when the lines A11 to A15 are at logic 0 will be repeated for every other possible combination of signals on these lines A11 to A15.

This is harmless if only the two memory chips are in use and provided the programmer is aware that an address such as 0000101011010011 will fetch the same data as 0000001011010011. For some purposes, particularly if some higher address numbers are used for other purposes such as accessing a keyboard, it may be necessary to have *complete* address decoding, so that the memory chips (continuing the example of Fig. 4.5) are activated only when the voltages on the lines A11 to A15 are all low.

This can be done by gating these lines so that a 1 on any line will produce a signal that disables both of the memory chips — in this

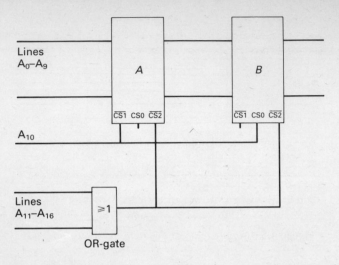

Figure 4.6   *A complete decoding system which will ensure that the memory chips are used only when lines A11 to A16 are all at zero.*

example, a 5-input OR gate system (Fig. 4.6) would provide the gating we need. More elaborate gating, such as when one particular address (for a port usually) has to be reserved, can be obtained by using chips known as decoders or demultiplexers — the address lines are connected to one set of pins on a decoder chip, and a select/enable signal is available at another pin. Designers of microprocessor systems often make use of a mixture of gates and decoder chips to obtain the address decoding that they require for any particular circuit.

Summary 4.2

Address decoding is the circuit-gating technique which is used to allow different sets of memory chips to respond to different groups of address numbers. The gating is carried out using the signals on the higher address lines (which are not connected to the address pins of the memory chips), and it requires chip select/enable pins to be provided on the memory chips.

Several chip select/enable pins are usually provided on a memory chip, so as to reduce the amount of external gating that is needed. For simple circuits, only partial decoding may be needed, in which the address numbers of the same set of chips repeat as the program counter increments, but complete address decoding will be needed if a port (for example) has to respond to one particular address and one only.

Exercises 4.2

1. What is address decoding?
2. Give an example of the use of address decoding.
3. What pin functions must be present on memory chips if address decoding is to be used?
4. What signals are used to carry out address decoding?
5. What would pins labelled as CS0, $\overline{\text{CS1}}$ be used for on a memory chip?
6. Why are several chip select/enable pins often used on each chip?
7. What is partial address decoding?
8. How is it possible to have the same memory activated by two different address numbers?
9. When is complete address decoding essential?
10. What is a decoder chip used for?

Read/Write Control

All of our discussion of memory so far assumes that the memory is being read only, and so the description of memory organization and address decoding will apply to ROM. For RAM, however, there is the added complication that the memory can be written as well as read, so that some way of controlling the actions of reading and writing is needed. When reading is enabled, the output of each memory cell that has been selected by the address and the chip select will be connected to the data pins of the chip, but when writing is selected, it will be the *input* of each selected memory cell which is so connected, because writing requires that the signals put on to the data lines by the microprocessor shall be copied into the memory cells.

This is achieved by using a read/write pin (Fig. 4.7), whose logic voltage (0 or 1) will decide which of the two operations is to be carried out. The conventional method is to use logic 1 for reading, logic 0 for writing. This, of course, entails an additional signal which must be sent out from the microprocessor, the Read/Write signal. This is one of a set of control signals which are available from the microprocessor, and which are connected along lines called the control bus. Though all

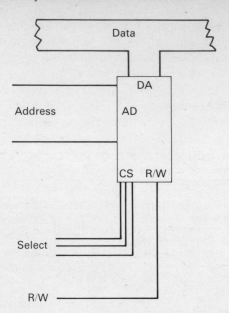

Figure 4.7 *The complete set of connections for a RAM chip — address, data, select, and read/write.*

microprocessors of similar capabilities (we cannot compare 8-bit microprocessors with 16-bit microprocessors) use similar arrangements of data pins and address pins, the control pins differ considerably from one microprocessor design to another. The Read/Write control pin is, however, used in one form or another in all types of microprocessor.

Data in Memory

The data bits which are present in a ROM chip are fixed; each one is a connection to logic 0 or logic 1 which is created at the time the chip is manufactured and which cannot be altered (except destructively) afterwards. The bits stored in RAM are volatile, meaning that they are not permanent but will all change to 0 when the power supply is switched from the chip. The bits can also be changed by writing new data to the chip.

The volatile nature of RAM can cause problems when a system containing RAM is switched on. If we regard each memory cell as a tiny electronic switch which can take up one of two possible positions, 1 or 0, with each one equally likely, then we can see that turning on power to these cells will cause some to switch on storing 1, some to switch on storing 0. This process is almost completely random, though tiny differences between cells may cause some cells always to switch on to 1, others always to switch on to 0. As a result, though, it appears that bytes are present at each address just after the power has been switched on; the memory is *not* filled with 00 bytes, as might have been expected. These bytes which appear at random do not represent a true sequence of instructions or data, and are called 'garbage'. The danger is that this garbage can be mistaken for a program, since it is never easy to distinguish the two. If the microprocessor is allowed to read the garbage, however, the system will 'crash'. This does not mean any electrical or physical damage, it simply means that the microprocessor will be executing a set of meaningless instructions at random, as a result of which it will probably write more garbage into the memory. The problem is that if a genuine program is stored anywhere in the RAM, it will probably be 'corrupted' (have its bytes changed) by the runaway microprocessor. In addition, the garbage program will not stop — some way of interrupting the action of the microprocessor will have to be used to restore control.

Computer systems which use microprocessors very often make use of a short piece of program in the monitor which eliminates garbage by writing bytes to all the RAM addresses. A popular method used in computers which make use of the Z-80 microprocessor is to write the sequence 00 FF 00 FF 00 FF... (hex) into consecutive memory addresses, and then read this back to check that the memory is operating correctly. This scheme is particularly suitable for Z-80 based computers, because if the microprocessor reads the byte 00 it carries out no action, and reading FF can be arranged to cause a return to the monitor, so that when the memory is filled with these bytes, reading a piece of memory unintentionally will cause no problems.

Summary 4.3

The microprocessor controls the use of memory by means of control signals passing along a set of lines called the control bus. One very important control signal is the read/write (R/W) signal, which switches

RAM chips between the two functions of writing (storing data from the microprocessor) and reading (passing data to the microprocessor).

Because RAM memory is volatile, it clears when the power is switched off, and will store 'garbage' bytes when the power is switched on again. These garbage bytes have been produced by the random switching of the memory cells, and are not in any sequence that would make sense any more than we could expect letters drawn at random from a Scrabble box to make meaningful sentences.

Exercises 4.3

1. What extra signal is needed by RAM as compared to ROM?
2. From where does this extra signal originate?
3. What is the name of the bus which carries such signals?
4. What is the meaning of 'volatile' as applied to memory?
5. What happens to the bits stored in a volatile memory when the power is switched off?
6. What happens to the cells of a volatile memory when the power is switched on?
7. What is meant by 'garbage'?
8. What is a program 'crash'?
9. What is a corrupted program, and how can this happen?
10. How can 'garbage' be eliminated?

End-of-Chapter Test

1. Explain the terms ROM and RAM.
2. Explain the difference between 'read' and 'write' operations.
3. Name the two types of RAM which are available.
4. What circuitry is needed in a memory chip apart from the memory cells themselves?
5. What is meant by a 2048 × 4 memory?
6. How would 8K × 2 chips be organized into a 8 Kbyte memory system?
7. What is meant by (a) partial (b) complete address decoding?
8. Why must each address number be unique?
9. What maximum address number can be obtained using 16 address lines? Answer in denary and in hex.

10. Explain the 'refreshing' of dynamic memory.
11. The upper nibble of a byte is AH. Explain what this means.
12. Why do memory chips always include at least one pin labelled CS or CE?
13. How many memory lines are unused on a 16-line bus if the memory extends from 0000H to 2FFFH?
14. How does the microprocessor separate the actions of reading and writing, when both commands start by placing an address on the address bus?
15. What problems are caused by the use of volatile memory?

5

Instructions

Instruction Sets

THE instruction set of a microprocessor is a list of all the instructions which that microprocessor type can carry out, along with the code number/s (in hex) for each variety of instruction. A full instruction-set will show clearly the result of each instruction, the effect on each register, particularly the status register, along with the time in clock cycles that is needed to carry out each instruction. This makes a full instruction set for a modern microprocessor a very lengthy document — my instruction set for the Z-80 is a book of some 270 pages — so that abbreviated instruction sets are very useful. These show the instruction in shortened form, called mnemonics (pronounced nemoniks), and the action indicated by symbols, and can usually be reduced to the size of a card that can be carried in a pocket.

Instruction Groups

All of the instructions in the instruction set of a modern microprocessor can be placed in one of three main groups. These are:

      (a) the data-transfer group,
      (b) the arithmetic and logic group,
      (c) the test and branch group.

The relative numbers of instructions in each of these groups depends very greatly on the design of the microprocessor, but all microprocessors use instructions which fit these groups, though some instructions may have to be placed in a fourth miscellaneous group. The data-transfer group consists of instructions which, as the name suggests, carry out the transfer of data bytes from one location to another. They are very important commands because most of the time that is needed for a program to run consists of time used in transferring data. Designers of microprocessor chips are well aware of this, and much ingenuity has gone into methods of making data-transfer operations as fast as possible, and with the minimum use of memory. The fastest of these data-transfer instructions are those which copy a data byte from one register to another, because these operations are completely internal to the microprocessor — no addresses or control signals need to be sent out, nor further data read in so as to complete these instructions. The number of such instructions depends very much on the number of registers that the microprocessor uses, and the design principles of the microprocessor — the 6502 uses no register-to-register transfers, for example, whereas the Z-80 uses 57 single-byte register transfer instructions.

EX DE, HL

Operation: DE ↔ HL

Format:

Opcode	Operands
EX	DE,HL
11101011 EB	

Description:
The two-byte contents of register pairs DE and HL are exchanged.

M CYCLES: 1   T STATES: 4   4 MHz E.T.: 1.00

Condition bits affected: None

Example:

If the content of register pair DE is the number 2822H and the content of the register pair HL is the number 499AH, after the instruction:

EX, DE, HL

the content of register pair DE will be 499AH and the content of register pair HL will be 2822H.

(Reproduced by kind permission of SGS-ATES Ltd)

Figure 5.1 *A sample entry from a full (Z-80) instruction set.*

LDA STOR	– load accumulator from memory address	(6502)
STA STOR	– store accumulator to memory address	(6502)
LDX STOR	– load X register from memory address	(6502)
LD A, (STOR)	– load accumulator from address	(Z–80)
LD (STOR),A	– store accumulator to memory address	(Z–80)
LD A,C	– load accumulator from C register	(Z–80)

Figure 5.2 *Some data transfer instruction (6502 and Z-80).*

The other important data-transfer operation is transfer between registers and external memory. Conventionally, a transfer (which is really a copying operation) of data from a memory location to a register is called a *load*, and a transfer from a memory location in the other direction is called a *store*, but the Z-80 instruction set calls both types of transfer *loads*, distinguishing the direction of transfer by the order of writing the names. For example, writing LD A, (STOR) means loading the accumulator from a memory (labelled STOR), whereas LD (STOR),A would mean that the byte from the accumulator was transferred to a store address (Fig. 5.1).

Data may also be transferred between microprocessor registers and the port chips which are used for input and output. If the port chip is memory-mapped, meaning that it is activated by decoding (fully) some memory address, then the data-transfer instructions that apply to the port are the same as would be used by any memory transfer. Some microprocessors have special provision for activating ports with a combination of address and control signals, so that commands such as IN or OUT can be used along with a single-byte address to control the transfer of data. This method has the advantage of needing much less decoding. Fig. 5.2 shows some examples of data-transfer instructions for a variety of microprocessor types.

Summary 5.1

The instruction set for a microprocessor is its software specification — a list of commands and their effects. These can generally be grouped as data-transfer, arithmetic/logic and test/branch commands. For all microprocessors, the data-transfer commands are of very great importance, because most of the time of a running program is spent in data transfer. Designers of microprocessors have endeavoured to make their data-transfer instructions as efficient in terms of time and memory-use as possible.

Exercises 5.1

1. What is an instruction set?
2. Why is a full instruction set so lengthy?
3. What are mnemonics?
4. Name the three main groups of instructions.
5. Into what instruction group would you place a 'store accumulator contents into memory' instruction?
6. Into what instruction group would you place a decision step?
7. Into what group would you place a *rotate* instruction?
8. Which of the data-transfer instructions are fastest performed?
9. Explain the terms *load, store*.
10. What is meant when a port chip is said to be 'memory-mapped'?

Arithmetic and Logic

The arithmetic and logic group of instructions consists mainly of the add and subtract instructions, along with the logic gating, shift and rotate instructions. More complex arithmetic, such as multiplication and division, is not generally available in the instruction sets of the smaller 8-bit microprocessors, though it is common on 16-bit microprocessors and available on the 8-bit 6809.

Addition and subtraction make use of the carry bit, stored in the status register, and used to enable numbers of more than the capacity of the accumulator to be added or subtracted. The usual design method is to add the carry bit automatically into any addition, clearing the carry (resetting the carry bit to 0) after the first bit of the addition, and setting it again if another carry is generated. Because of this, if two separate addition instructions are carried out, one following the other in a program, there is a possibility that the carry may be set by the first addition and so added, unwanted, into the second. Novice programmers often find it hard to explain why 1 + 1 = 3 in such cases! In microprocessors which use this method, the carry bit should always be reset before an addition unless a carry from a previous addition is wanted. The alternative (used by the Z-80) is to have two sets of arithmetic instructions, one set of which makes use of the carry, the other of which does not. This second method is easier to remember, but makes the instruction set larger. Fig. 5.3 shows examples of both techniques.

ADC	– add with carry	(6502)
CLC	– clear carry bit	(6502)
SEC	– set carry bit	(6502)
ADC A,D	– add with carry content of D to A	(Z–80)
ADD A,E	– add, no carry, content of E to A	(Z–80)

Figure 5.3 *Some arithmetic instructions (6502 and Z-80)*

An instruction which fits into this group, but which has not been considered earlier, is the *compare* instruction. The *compare* action is similar to that of subtraction, but it does not give the result of the subtraction in the accumulator register, only in the status register. For example, if the accumulator contained 0FH, and the instruction COMPARE 0FH were executed, the accumulator would still contain 0FH at the end of the instruction, but the zero flag in the status register would be set to indicate that the *compare* action had produced a zero result. The *compare* action is used to a considerable extent in test-and-branch actions (see p. 72).

The logic part of this group contains the AND, OR, and XOR instructions. When these instructions are being used, each bit of a byte is considered separately and individually, not as part of a number. Each of the three logic actions consists of a comparison between bits in corresponding places in the two bytes. The rules for the logic actions were covered in Chapter Two, but are recalled in truth-table form in Fig. 5.4. An 8-bit microprocessor will carry out these actions on eight bits of a byte at a time, comparing the bits of two bytes in pairs with each other. Since the logic action is purely a comparison of individual bits, there can never be a carry generated nor used by any of these logic actions. The usual sequence is that a byte stored in memory is gated with another stored in a register, and the result is stored in the same register, usually the accumulator. The shift and rotate actions, though belonging to the arithmetic and logic group, require no additional bytes, because they are operations that are carried out on a single byte (considering an 8-bit microprocessor) only. These actions also were explained in

Chapter Two and are recalled in Fig. 5.5. The importance of shifting is that in arithmetic, a shift can give the result of a multiplication or division by two. Either shifting or rotation can also be used as a method of feeding a byte out of a register one bit at a time, as is needed to communicate between one computer and another over telephone lines or radio links. Fig. 5.6 shows some examples of commands.

AND				OR				XOR		
A	B	Q		A	B	Q		A	B	Q
0	0	0		0	0	0		0	0	0
0	1	0		0	1	1		0	1	1
1	0	0		1	0	1		1	0	1
1	1	1		1	1	1		1	1	0

Figure 5.4 *The logic truth tables recalled.*

Shift	Rotate
11110000	11110000
shift left	rotate left
11100000	11100001
11110000	11110000
shift right	rotate right
01111000	01111000

Figure 5.5 *Shift and rotate instructions recalled.*

AND	– AND byte from memory with accumulator	(6502)
ORA	– OR byte from memory with accumulator	(6502)
EOR	– XOR byte from memory with accumulator	(6502)
ASL	– shift left, using carry	(6502)
ROR	– rotate right, using carry	(6502)
AND B	– AND byte in B with byte in A	(Z-80)
OR C	– OR byte in C with byte in A	(Z-80)
XOR L	– XOR byte in L with byte in A	(Z-80)
SLA D	– shift byte in D left, using carry	(Z-80)
RR H	– rotate byte in H right, using carry	(Z-80)

Figure 5.6 *Some logic commands of the 6502 and Z-80.*

Test and Branch

The third very important group of instructions consists of the test and branch (or jump) operations which carry out the decision steps of a flowchart. The 'test' which is made consists of a test of one of the 'flags' of the status register, and if the test succeeds (the flag bit is in the state that is being tested for), then a branch or jump is made. This means that the program counter is forced to store a different address, and the program will proceed from this new address, using the normal incrementing action of the PC. If the test fails, however, the new address is not used, and the program counter increments in the normal way to bring up the next address in sequence.

Each test can have only two possible results — pass or fail. The three most important types of test are for the carry flag, the zero flag and the sign flag. The carry flag of the status register will be set (to logic 1) if an arithmetic operation has resulted in a carry out of the most significant place of the byte, otherwise it will be reset (to 0). The instructions: 'jump if carry set' and 'jump if carry reset' can both be used so that a jump to a new address can be forced if either condition is true. The choice makes it easier to program, as you will find when further work on flowcharting is covered later in this book.

Similarly, the zero flag of the status register will be set if an action has caused a zero result, and jumping or branching can be arranged for the two possible conditions of zero flag set or zero flag reset. One very common use of this flag is when a program is arranged to detect the presence of a specific byte, such as 0DH. By subtracting this (or using the compare instruction) from a byte in the accumulator, the zero flag will be set if the accumulator content matches the byte being looked for, and the jump can be taken.

Finally, the sign flag can be used to control a jump or branch either for a positive result or for a negative result. A negative result is interpreted as being a binary number with its most significant bit set (logic 1); a positive result as meaning that the m.s.b. is 0. Fig. 5.7 shows some examples of test and branch instructions.

Not all jump/branch instructions are conditional on the status flags. There are unconditional jump/branch instructions, which will cause a change in the address in the PC, and hence on the address bus lines, no matter what flags in the status register are set or reset. In addition, most microprocessors include the instructions CALL and RET (a short form of return) which cause a different type of branching. The CALL instruction has to be followed by a memory address, and its effect is to

BCS	– branch if carry set	(6502)
BEQ	– branch if Z-flag = 1	(6502)
BMI	– branch if N-flag = 1	(6502)
JR Z	– jump if zero-flag set	(Z–80)
JR C	– jump if carry set	(Z–80)
JP M	– jump if sign negative	(Z–80)

Figure 5.7   *Some test-and-branch commands of the 6502 and Z-80.*

JSR	– jump to subroutine	(6502)
RTS	– return from subroutine	(6502)
CALL	– call subroutine	(Z–80)
RET	– return from subroutine	(Z–80)

Figure 5.8   *Call and Return commands of the 6502 and the Z-80.*

cause a program which starts at that address to be executed.

This is normally a temporary action, and the piece of program which is being executed as a result of the CALL instruction will end with a RET instruction, which has the effect of returning to the main program, at an address just following the CALL instruction. The instruction commands are illustrated in Fig. 5.8. CALL and RET are used so that a piece of program, called a *subroutine,* which is needed many times in the course of a program, can be called from any part of that program. For example, a system might have to display numbers in ASCII form each time a result was obtained for a calculation. Rather than have a program for carrying out the conversion of binary to denary to ASCII at each point where it is needed, one such program is written, ending with a RET command, and is placed in a different part of memory. Each time this program is needed, it can be used by the instruction CALL followed by its starting address. When the program has done its work, the RET instruction will cause a return to the section of the main program immediately following the CALL instruction.

Summary 5.2

The arithmetic/logic group consists of the add, subtract, compare, shift and rotate and logic gating instruction. Of these, only the add, subtract and compare affect the carry bit. The logic actions affect the separate bits of a byte rather than treating the byte as a number, and the gate actions can be summarized in truth tables. Shifting and rotating are actions that are carried out on one byte, unlike the other actions of this group which affect two bytes each.

Testing and branching are the actions which enable the microprocessor to carry out the decision steps of a flowchart. These instructions cause the microprocessor to jump (branch) to a new program address depending on the result of a test made on the flag bits in the status register. The flags which are most often used for test purposes are the carry, zero and sign flags. Unconditional branch instructions are also used. CALL and RET are commands which are used so as to permit a section of program, called a *subroutine*, to be used in several places in a program, though it is located in a single part of memory.

Exercises 5.2

1. List the main instruction of the arithmetic and logic group.
2. Which instructions make use of the carry bit?
3. Which instructions act on one bit only?
4. Where is the result of an XOR process stored?
5. Draw truth tables for AND, OR and XOR actions.
6. What is the result of ANDing 3BH with E2H?
7. What is the result of shifting F2H left twice?
8. Why are test and branch commands important?
9. What register must be used by test and branch commands?
10. What are the three conditions most often tested for?

Addressing Methods

Many of the instructions in the instruction set of a microprocessor require a byte to be fetched from memory to carry out the action of the instruction, or for an address to be specified to store a byte. During the execution of these instructions, an address will have to be put on to the address lines, and if we assume that the microprocessor is an 8-bit one, then the address will consist of two bytes. This address, sometimes called the *effective address* (EA) for the instruction, must be read into the microprocessor from the program memory, and the convention is that the address data will always follow the command byte or bytes. Some commands, of course, do not need any bytes read in following the instruction byte. A shift operation, for example, will use only the byte that is already present in a register; or an instruction such as *clear carry* may change only one bit in one register. These types of operations are sometimes said to use implied addressing, meaning that the address of the byte or bit is implied in the code of the instruction, and nothing more is needed.

Immediate Addressing

Most commands, however, need some sort of additional address data to be fetched following the command byte, and the different address methods that are used are responsible for the large instruction sets of so many modern designs of microprocessors. The simplest way of putting data into a program is by placing the data bytes in the action program memory itself, and this is called *immediate addressing.* When a command uses immediate addressing, the byte which is to be used is stored in program memory, immediately following the instruction code, hence the name. For example, if we had an instruction: load-immediate, 0FH, then the load-immediate code would be stored in program memory, and the byte at the next higher address would be 0FH, the byte which is to be loaded. (Fig 5.9). Immediate addressing is unique in as much as it needs no address data to follow the instruction. It is, however, a very limited method, because the data which are used in this way are part of the program, and there is no store-immediate command (though there are versions of all the other main commands for immediate addressing).

LDA #34H (load the byte 34H)
Bytes in program memory: A9 34
(6502)
LD A, 34H (load the byte 34H)
Bytes in program memory: 3E 34
(Z–80)

Figure 5.9 *Examples of immediate loading instructions.*

When an address which is not the next address in program memory has to be specified, the most obvious method of so doing is known as extended addressing; it is the method that was used in the trace table of Fig. 3.12. When extended addressing is used, each command byte is followed by two address bytes, giving (for an 8-bit microprocessor) the full address in memory of the byte which is to be fetched. Conventionally, these address bytes are stored with the lower byte of the address fetched first, so that if we assume that a byte is to be fetched from location 7F2BH, and the load command code is DAH, then the sequence of storing the bytes in program memory is DA 2B 7F. An example is shown in Fig. 5.10.

LDA 4217H (load A from address 4217H)

Bytes in program memory: AD 17 42

(6502)

LD A,(4217) (load A from address 4217H)

Bytes in program memory: 3A 17 42

(Z–80)

Figure 5.10   *Examples of extended (or absolute) addressing.*

Extended Addressing

Extended addressing is comparatively simple, but its use is time and memory consuming. It is time consuming because two fetch cycles have to be used after the instruction byte has been decoded, and the complete address has to be assembled before the address can be put on the bus. It is memory consuming because two bytes of program memory are needed to store the address, as well as one byte of RAM for the data. One further disadvantage, which is more important when advanced programming is considered, is that the address must be known to the programmer when the program is written. In some types of program (especially these involving printing data on a VDU screen) it is necessary to use addresses which have been calculated by the program itself, and extended addressing does not lend itself so well to this type of use as some other methods.

Designers of microprocessors have therefore devoted much time, ingenuity, and effort into devising other ways of obtaining addresses.

Some of these methods are simple and useful for elementary programming, others are complex and useful only to the advanced programmer.

PC-relative addressing is a method which was used to a considerable extent in some early microprocessor designs, but which nowadays is used only for some jump/branch types of instructions, sometimes for *call* instructions. An instruction which is PC-relative addressed is arranged so that the instruction byte is followed in the program memory by a single byte, called a *displacement*. When this displacement byte is fetched from program memory, it is added to the *lower* byte of the address in the program counter register, with any carry discarded, and the resulting address is put on to the address bus lines.

> BNE 12H (branch if Z=0, jumping over 12H bytes in program memory to higher address number in PC). (6502)
>
> JR NC,08H (jump over 8 bytes to higher address number if carry bit is not set). (Z–80)
>
> Note that modern microprocessors use this method only for branch (jump-relative) instructions.

Figure 5.11 *Examples of program-counter relative addressing.*

Using a single byte in this way allows us, as Fig. 5.11 shows, to move to an address either ahead of or behind the PC address at which the instruction occurs. The maximum displacement extends to 127 (denary) steps forward from the current PC address, or 128 (denary) steps backward, but for the type of instruction for which PC-relative addressing is used, this is seldom a handicap. Most beginners to programming usually wish that they could write programs as long as 128 steps altogether!

A variation on PC-relative addressing is indexed addressing. The meaning of indexed addressing is differently interpreted by different manufacturers of microprocessors, but the usual meaning is that the displacement byte which follows the instruction code is added to the lower byte of an address stored in a special register, the index register, with the carry discarded as before. This resulting address is placed on the address bus lines to form the effective address for the byte that is to be fetched (Fig. 5.12).

Zero-page addressing is another method which makes use of a single byte read in from program memory following the instruction byte. When the zero-page form of an instruction is used, the upper byte of the

Note that the use of indexing in the 6502 is quite different to the method used in the Z–80.

LDA (307F), Y—load A from the address found by adding the number in the Y–register to the address 307F.

(6502; the Y–register is a single-byte register.)

LD A(IX + 5)—load A from the address stored in the IX register + 5 (Z–80; the IX register is a two-byte register.)

Figure 5.12   *Examples of indexed addressing.*

effective address is assumed to be zero (hence the name), and the byte that is read in from program memory is taken as being the lower byte of the address. If we assume, for example, that 9AH is the instruction byte for LOAD (with zero-page addressing), then the sequence 9A 2D in program memory would cause a byte to be loaded from the address 002DH. Zero-page addressing is used to a very considerable extent in the 6502, to a lesser extent in the Z-80, and a variation of the method, which allows the upper byte to be chosen for a number of instructions, is used on the 6809.

Indirect Addressing

The last method we shall look at here is a variety of an important addressing method, which is called *indirect* addressing. This implies that the byte or bytes which are stored following the instruction code in program memory do not lead directly to the effective address, and this is one form of addressing which permits an address which is not known to the programmer at the time of writing the program to be used, since the address does not have to be contained within the program in any way.

The simplest type of indirect addressing is called register indirect, and is a method which is used extensively in the Z-80. In a register-indirect load, the address which is used and placed on the address lines is contained in a 16-bit register. The Z-80 has three such registers (actually pairs of 8-bit registers), and only a single-byte instruction is needed to load a byte into the accumulator from an address held in one of these register pairs (Fig. 5.13). The register pairs can be loaded with the address during the course of the program, so obtaining address bytes

LD HL,4013H register-pair HL loaded with address 4013H

LD A, (HL) load A from address in HL

INC HL increment HL address to 4014H

LD B, (HL) load B register from address in HL

(All from Z–80.)

The 6502 indirect-addressing methods are of different and less simple types.

Figure 5.13   *Example of register-indirect addressing (Z-80).*

that have been derived by the program itself and which could not have been known to the programmer, though the programmer must program the method by which the address bytes are to be obtained.

Summary 5.3

Most commands require one or more bytes to be read from memory in addition to the instruction byte, and an address for each such byte (the effective address) must be put on to the address lines. Methods for obtaining this address are called the addressing methods for the microprocessor.

Immediate addressing locates the byte that is to be used by the instruction in the program memory immediately following the instruction byte. This is very efficient, but if the program is fixed in ROM it prevents this byte from ever being changed.

Extended addressing requires two bytes following the instruction byte in the program memory; these bytes are then assembled into a two-byte address. This is simple, but time and memory consuming. Many other forms of addressing have been developed to try to reduce the time that is needed to fetch data, to make efficient use of memory, and to allow address numbers to be changed in the course of the program. Relative addressing makes use of a 'displacement' byte which is added to the lower byte of either the PC address or an address held in an 'index' register. Zero-page addressing uses only one address byte, the lower byte, assuming that the upper byte is 00. Register-indirect addressing uses an address number stored in a register — this address can be easily changed in the course of a program.

Exercises 5.3

1. What is an 'addressing method'?
2. Give an example of a command which does not need an address.
3. Describe immediate addressing. What is its main disadvantage?
4. Describe extended addressing. Why does it need a lot of time and memory?
5. What is meant by PC-relative addressing?
6. What is the name given to the byte in the program memory that has to be read to form a PC-relative address?
7. What are the limits to the number that can be obtained for a PC-relative address?
8. What is zero-page addressing?
9. What is meant by indirect addressing?
10. What is register-indirect addressing?

End-of-Chapter Test

1. Why is the instruction set of a microprocessor called the 'software specification'?
2. What facts would be presented in an abbreviated instruction set?
3. Name the three main groups of instructions.
4. Why can some instructions be performed much more quickly than others?
5. List the main instructions of the arithmetic and logic group.
6. Which of the instructions of Question 5 need two bytes to act on?
7. Which of the instructions in Question 5 do *not* generate a carry bit?
8. Name and write truth tables for the three logic gate actions.
9. What main tests can be made in a test and branch instruction?
10. How are the CALL and RETurn instructions used?
11. What is meant by effective address?
12. Explain and illustrate immediate addressing.
13. What advantages does extended addressing have compared with immediate addressing?
14. Why is zero-page addressing faster than extended addressing?
15. Why should indirect addressing be important?

6
Looping Programs

Branching and Looping

As defined in Chapter Three, a branching program permits more than one path to exist between the START and the END marks of its flowchart. An extension of this is a looping program, in which the branch leads back to an earlier section of the program, causing a number of steps to be repeated until finally a test step ceases to cause the branch to be taken. The loop is the section of program which is repeated, and each execution of the steps of the loop is called a 'pass' of the loop. Loops may be holding loops, counting loops or searching loops. In a holding loop, the quantity which is tested at the branchpoint in each pass of the loop will be constant for some time, so holding the loop, and then will change, causing the program to break out of the loop. In the counting loop, as the name suggests, a quantity is counted by being incremented or more often decremented, on each pass through the loop until it reaches a predetermined amount, usually zero. In a searching loop, a byte is read in on each pass through the loop and compared with a fixed byte until a match is found.

A holding loop can be used, for example, to wait for a particular key on a keyboard being pressed, a counting loop can be used to ensure that a certain number of bytes are shifted from one part of memory to another, or for time delays; and a searching loop can be used to find the end of a set of bytes in memory. Fig. 6.1 shows two flowcharts which illustrate elementary versions of the first two types of loops. Fig. 6.1(a) shows a simple holding loop. When the loop is entered, the accumulator

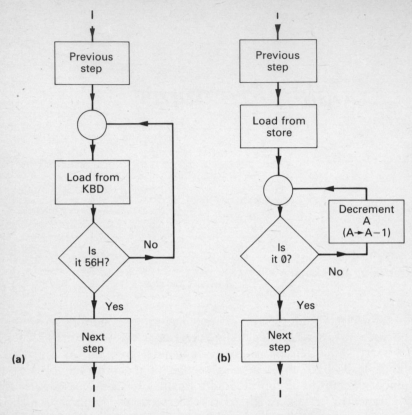

Figure 6.1 *Loop flowcharts. (a) A holding loop which awaits the entry of a particular byte (56H in this example). (b) A counting loop which loads a number from memory, and decrements it in each loop until it is zero.*

is loaded from the location KBD (a keyboard) and the next step tests for the byte 56H, which is the ASCII code for the letter *V*. This is the code byte which will come from the keyboard if the !etter *V* is pressed, and if it is not, then the test fails and the program loops back to the load-A-from KBD step again. The microprocessor will stay in this loop until the *V* is pressed or until the system is reset or switched off.

Fig. 6.1(b) shows a simple counting loop used in this example as a time delay. When the loop is entered, the accumulator is loaded with a byte from memory (STORE), and the decision step tests for this byte being zero. If it is not, then the byte is decremented (subtracting 1), and the test is carried out again. Until the test has been carried out as many times as is needed by the number in STORE, the decision step will cause

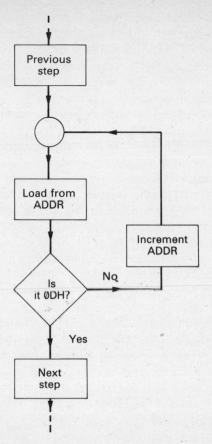

Figure 6.2 *A searching loop which compares a byte taken from memory with a fixed byte (0DH in this example). If the bytes do not match, the memory address is incremented, and the process repeated until a match is found.*

the loop to be repeated. Since the test and decrement steps take time to perform, this is a simple way of creating a time delay for programs which rely on timing, such as clock programs or programs which feed bits in or out at a definite rate (cassette data recording and replay, using printers, etc.).

Searching loops are extensively used in computing. Fig. 6.2 shows a flowchart which causes the microprocessor to examine a set of addresses, looking for the byte 0DH. This is used as the code for a 'carriage return' (end of line) in ASCII code, so that the ability to find this byte in a long string of bytes allows lines of coded characters to be sorted out. When the loop is entered, ADDR is the address in memory of the start of a set of code bytes. The first instruction of the loop consists of loading the byte at address ADDR into the accumulator. The byte is then tested. If it is not equal to 0DH, then the address number is incremented, so that ADDR now is a number which is one greater than its previous value. The test is now repeated by looping back to the load instruction, so that the next address is used to load the accumulator. The sequence of incrementing, loading and testing is continued until a match is found.

The searching loop is particularly easy to implement using microprocessors with register-indirect addressing, such as the Z-80, and typical Z-80 instructions for such a loop are shown in Fig. 6.3(a), with explanations. By holding the starting address in the register pair HL (High byte-Low byte), the incrementing action can be carried out by using the single-byte instruction INC HL, making the programming particularly neat and simple. The 6502 microprocessor type can tackle this problem by using a form of indexing, adding a byte from a register to the 'base address', and incrementing the number in the register at each pass through the loop—Fig. 6.3(b) shows this quite different technique.

In a searching loop, the operand of an instruction is being changed on each pass through the loop. The 'load A from address' instruction in Fig. 6.3(b) uses as its operand an address, and this address is being incremented each time the loop is used; the method used in Fig. 6.3(a) changes the operand indirectly, so that the same loading instruction is being used, but the *content* of HL is incremented. In a counting loop, it is the contents of memory or a register which is changed on each pass through the loop, not the memory address or the address number in a register. Normally, the content of the memory or register that is used in a counting loop will be a number that is decremented to zero in the counting loop, because zero can be tested for automatically by the use of the zero flag in the status register.

Summary 6.1

A branching program permits more than one path to exist between the start and the end of a program, a looping program allows a section of program (the loop) to be repeated several times, each repetition being called a 'pass'. Loops may be holding loops, counting loops, or searching loops. A holding loop is used to test for an awaited byte from outside the system (a key being pressed, for example). A counting loop is used for a count or time delay, decrementing the value of a number on each pass through the loop. A searching loop is used to look at a sequence of addresses in memory for a given byte or pattern of bytes.

In some loops, searching loops in particular, the operand of one of the instructions may be changed at each pass through the loop. In a counting loop, the contents of a register or a memory will be changed, usually decremented, but in a holding loop there is generally no change made to any quantity in the course of the loop instructions.

(a)

```
          LD HL, START     ;  start addr.
LOOP :    LDA, (HL)        ;  load A from this address
          INC HL           ;  registers now hold next address
          CP 0DH           ;  compare byte in A with 0DH
          JR NZ, LOOP      ;  back if no match
          ...
          (next part of program)
```

(b)

```
          LDY #−1H         ;  store −1 in Y−register
LOOP :    INY              ;  increment Y (to 0 on first pass)
          LDA (START),Y    ;  load A from START + Y address
          CMP # 0DH        ;  compare 0DH
          BNE LOOP         ;  back if no match
          ...
          (next part of program)
```

Note: the hashmark (#) means immediate loading in 6502 code.

Figure 6.3 *Searching-loop program methods.* **(a)** *With Z-80 code, register-indirect loading can be used. The address held in the HL register pair is incremented on each pass of the loop.* **(b)** *With 6502 code, indexing is used. The index register Y is loaded with −1 at the start, then incremented to 0. The accumulator is loaded by the instruction LDA (START),Y (which means using the address START + Y). With Y = 0, this means using START on the first pass through the loop, then START + 1, START + 2, and so on.*

Exercises 6.1

1. What is the effect of a branch step in a program?
2. What is meant by a loop?
3. Name the three main types of loops.
4. Draw a flowchart of a loop which continues until the byte 44H is read in from location KBD.
5. What type of loop requires a number to be decremented on each pass?
6. What is the main feature of a searching loop?
7. A counting loop does not generally *increment* the count number in each pass. Why not?
8. What determines the number of passes of a holding loop that will be executed?
9. In what application of microprocessors are searching loops mostly used?
10. Why should time delays be useful in a program?

The Branch Step

The branch step of a loop must contain an operand which allows the next instruction to be located if the branch is taken. If the branch is not taken, the next instruction will be located simply by incrementing the program counter, but if the branch is taken, the address must be changed to some different value. The loop shown in Fig. 6.1(a) is the easiest type to program, because when the loop is taken, the path is directly back to some previous address. If a PC-relative addressing method is used, as frequently happens, the instruction code can be followed by a single byte (displacement) which will take the program back to the correct address if the branch is taken.

The addressing method for the program in Fig. 6.1(b) is equally straightforward, but the programming is not quite so simple. After the branch has been taken, the loop, as shown, is not immediately back to the previous piece of program, but to an additional step, the decrement step. This can be avoided, saving a lot of effort, by placing the decrement step in the main part of the program, as shown in the flowchart of Fig. 6.4, and this is the method that would normally be used. For the program of Fig. 6.2, in which an address is incremented, such a method is equally possible, by incrementing the address before testing the byte that has been loaded into the accumulator from that address.

Consider now the section of flowchart which is shown in Fig. 6.5. At the point where we join the flowchart, a byte is read from the memory and tested. The test is that the byte is less than 5BH, which if the memory contained a set of ASCII codes for letters of the alphabet, would correspond to an upper-case (capital) letter. If the byte is *not* less than 5BH (a lower-case letter), then the program proceeds to the next step, but if the byte is less than 5BH, the number 20H is added — making the ASCII code into the one for the corresponding lower-case

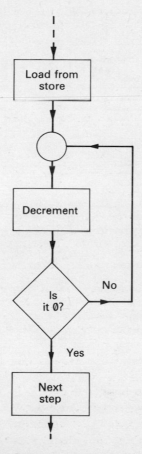

Figure 6.4 *The flowchart of Fig. 6.1(b) can be rearranged so that the decrement step is in the main section of program, not part of a jump to a different address. This makes programming much simpler.*

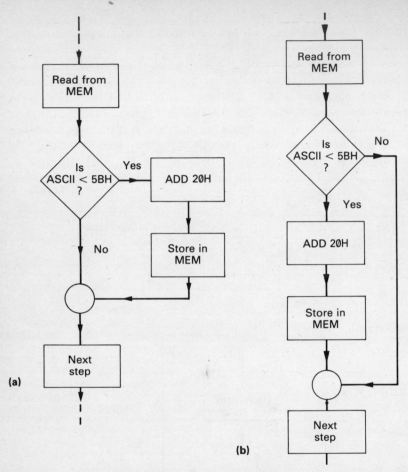

Figure 6.5 *More loop rearrangement.* (a) *A loop in which two actions are performed in the YES branch. This would probably be part of a larger loop in which bytes were read from successive memory addresses.* (b) *The loop rearranged to make programming easier.*

letter, so that all *A*'s are converted to *a*'s, *B*'s to *b*'s and so on. Note that this is only part of a flowchart, because we would need other steps to exclude some values of bytes (the codes for the digits 0 to 9, for example) from this treatment, but the principle is unchanged. The two steps which are in the Yes path of the decision step cannot be placed in any other path and must be carried out only on bytes whose values were less than 5BH when tested.

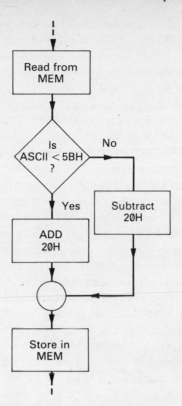

Figure 6.6 *A flowchart in which the action following the decision is completely different for the two branches.*

This flowchart can also be rewritten so that an easier jump is possible, simply by making the longer path, the path for the Yes decision, into the main one, and the short path, the route that is followed when the byte is greater than 5BH, into the jump step. The point of all this is that a flowchart can very often be improved so as to make a program much simpler by avoiding multiple-jump steps. If a decision step can be arranged so that one of its routes is directly back, with no other processing, to another step, then this should be done. When a decision step results in two different sections of program being executed, having nothing in common with each other, then more than one jump step will be needed for each decision step. Fig. 6.6 shows such a program section. This time, if the byte is more than 5BH, the 20H is subtracted; the rest of the program section is as before. Assuming that all of the other ASCII codes had been eliminated, so that this section of program dealt only with capital and lower-case letter codes, then the effect of this section would be to change all capitals to lower case, and all lower case to capitals.

Fig. 6.7 shows why the flowchart of Fig. 6.6 needs more than one jump step. This is a section of program which might be used to implement this portion of flowchart. The code that is shown is not specific to one microprocessor (though similar to Z-80), and the principles are usable with any microprocessor type. The first step is to load the byte from the memory address (coded as MEM) into the accumulator, so that it can be compared in the second step with 5BH. If the number which has been loaded from memory is greater than 5BH, then the carry flag will remain reset, but if the number is less than 5BH, the carry flag will be set because of the 'borrow' in the subtraction which the COMPARE instruction carries out. The jump step, JR NC,SUBIT is the abbreviated way of writing the instruction — if the carry flag is not set (NC), jump to a new piece of program at address labelled SUBIT.

If the carry flag is set (number is less than 5BH) then the next step is the fourth one — 20H is added to the number in the accumulator and in step 5, labelled MEMIT, this number is stored back in memory, replacing the number that was tested. Once this has been done, step 6 causes a jump to step 9, labelled NEXT, so skipping over steps 7 and 8 which are not wanted in this particular branch.

When the carry flag has not been set by the test, the program jumps from step 3 to step 7, at which step 20H is subtracted. So that this can now join the rest of the program again, the next step that must be executed is step 5, MEMIT, so following the path taken by the other branch of the program after the byte was altered. The program steps for the two different branches are: 1, 2, 3, 4, 5, 6, 9 if the carry bit is set, and 1, 2, 3, 7, 8, 5, 6, 9 if the carry bit is not set. Note the use of jump steps (a bridge) such as step 6 to prevent a piece of program from being executed in the wrong branch of the program.

Summary 6.2

A branch step consists of an operation code and an operand which will change the program counter address if the test succeeds and the branch is taken. The simplest type of branching (jumping) is to another section of the main program, and this is often possible, though the sequence of the flowchart may have to be slightly rearranged to permit it. In general, the longer of the alternative paths should be put into the main program

1.		LDA,(MEM)	; load A from MEM
2.		CP 5BH	; compare 5BH (a subtract action)
3.		JR, NC, SUBIT	; no carry means byte greater than 5BH
4.		ADD 20H	; byte <5BH, so add 20H
5.	MEMIT:	STA, (MEM)	; put it back in memory
6.		JR NEXT	; jump to NEXT
7.	SUBIT :	SUB 20H	; byte >5BH, so subtract 20H
8.		JR MEMIT	; jump to MEMIT to store byte
9.	NEXT:	... next part of program...	

Figure 6.7 *A program for the flowchart of Fig. 6.6. The instruction set is an imaginary one, but similar commands can be found in the instruction sets of modern microprocessors.*

to make this possible. If the actions that are carried out as part of the loop are not required in the main section of the program, then a second jump will be needed, and it may also be necessary to make a bridging jump over the section of program that is not executed if the branch is not taken.

Exercises 6.2

1. What is the form of a branch step?
2. How many bytes will follow the operating code if PC-relative addressing is used for a branch step?
3. Why does a flowchart sometimes have to be changed to make a branch step simpler?
4. What is meant by multiple jump (branch) steps?
5. Draw two flowcharts for a holding loop, and explain why one of them needs only one jump, but the other needs more than one jump.
6. Draw a flowchart for a section of program which will add 30H to each of a set of numbers read in from memory.
7. Draw a flowchart for a section of program which will read five bytes from memory and add 10H to each byte.
8. Draw a flowchart for a section of program which will read bytes consecutively from memory until the byte 0DH is read.
9. Draw a flowchart for a section of program which will read and count bytes from memory until 0DH is read.
10. Draw a flowchart for a program which will wait until the letter *K* is pressed on a keyboard (address KBRD) and then read bytes from memory until 0 is read, and then print (address PRNT) the number of bytes that have been read.

An Adding Program

This chapter concludes with an example of a looping program, starting with an algorithm, and ending with a program written for an (imaginary) microprocessor whose instruction set is shown in Fig. 6.8. The problem is to add together five single-byte numbers, which are stored in separate but consecutive memory addresses, and to store the result in a sixth address. For simplicity, we shall assume that the sum of all five numbers is less than 255D, or 0FFH, so that no carry will be generated and no additional storage needed.

We must start with a suitable algorithm, because there is no microprocessor instruction for adding five numbers. We could, of course, use a linear program with four ADD commands (one number loaded in, then the others added one by one), as shown in Fig. 6.9, but a repetitive action of this type seems ideally suited to a loop, using a counter byte to control the number of passes through the loop. A suitable looping algorithm would be as follows. Store the numbers in consecutive memory addresses, starting at an address which we can label as STOR. In this way, the first number is at address STOR, the next at STOR+1, the next at STOR+2 and so on. We can use whatever memory addresses we like when the program is finally written, so that STOR could be 42A6H, STOR+1, 42A7H and so on.

To carry out the action, we first place a counter byte 04H into an address labelled COUNT, then load the byte at STOR into the accumulator, and increment the memory address to STOR+1. The number at STOR+1 is now added into the accumulator, and the result is stored at the address labelled SUM. The memory address is now incremented to STOR+2, and the accumulator is loaded from the counter address (COUNT), and this number is decremented, and tested. If the count number is zero, the program halts (return to monitor), but if it is not, then the accumulator is loaded from SUM, so that it contains the sum of all the additions so far, and the program loops back to the addition step. The program will loop four times, and end with the sum of all five numbers placed in the address SUM.

Fig. 6.10 shows a flowchart drawn from this algorithm. When the program starts, the count is set by loading a memory address, COUNT, with 04H. The accumulator is then loaded from the address labelled STOR, which contains the first number, and the STOR address is incremented, so that when this address is used next time it will fetch the next number byte. This byte is then added in to the accumulator, and the sum stored, because the accumulator now has to be used to test the

Mnemonic	Meaning
LDA nn	Load accumulator with the number nn.
LDA,NN	Load accumulator from the address 00NN.
LDA,(NN)	Load accumulator from address stored in 00NN.
STA,NN	Store byte from accumulator to address 00NN.
ADD nn	Add immediate the byte nn.
ADD,NN	Add to accumulator the byte from 00NN.
ADD,(NN)	Add to accumulator the byte from address stored in 00NN.
DEC	Subtract 1 from number in accumulator.
INC	Add 1 to the number stored in accumulator.
INC(NN)	Increment the number stored in 00NN.
HLT	Stop program action.
CMP nn	Compare byte in accumulator with nn.
CMP,NN	Compare byte in accumulator with byte in 00NN.
JMP,NN	Jump to address 00NN.
JPZ,NN	Jump to 00NN if accumulator contains zero.
JNZ,NN	Jump to 00NN if accumulator not zero.

Figure 6.8 *The instruction set for SI-MPU-2. This simulation program allows microprocessor programs to be tested on a microcomputer. Note that nn is used to indicate immediate loading of a byte (using denary 0 to 255), and NN is used to indicate zero-page loading.*

```
LDA,STOR
ADD,STOR+1
ADD,STOR+2
ADD,STOR+3
ADD,STOR+4
STA,RESULT
```

Figure 6.9 *Five bytes can be added using a linear program, such as this.*

count. The count number is loaded into the accumulator, decremented, and tested. If the count number is now zero, the program has ended, but if not, the accumulator has the sum number loaded back, and the program loops back to the step at which the number in the memory address (which has been incremented) is added in to the accumulator. At the end of the program, the number in the address SUM will be the sum of all five numbers.

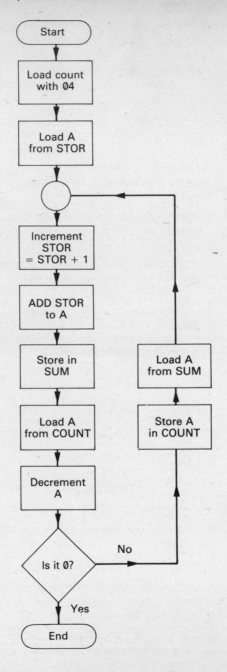

Figure 6.10 *A flowchart for adding five bytes, using a loop.*

A flowchart of the type shown in Fig. 6.10 needs some sort of indirect addressing, or indexed addressing, so that the loading can be from an address number that is incremented on each pass through the loop. The simple instruction set of the microprocessor in Fig. 6.8 shows three such instructions, two of which are LDA, (NN) and ADD, (NN). The meaning of the LDA, (NN) instruction is that the accumulator is to be loaded from an address that is contained in the byte which is stored at the address NN. The simple instruction set uses only page zero addresses, with denary numbers less than 256, so that if we imagine that the address NN is 100, then the instruction LDA, (NN) would first of all locate address 100, and find the byte stored there. If this byte were 64(D), then 64 would be the loading address, and the byte in address 64 would be the byte which would be loaded into the accumulator. The ADD, (NN) instruction operates in a similar way, using the address NN to locate another address which is the address of the byte that will be added to the accumulator. This scheme also makes use of another instruction, INC (NN), which means increment the number stored in address NN. This allows us to change the address so that, using the same example again, INC(100) would increment the address number (64D) in address 100, making it 65 so that the next time ADD(100) is used, the address from which the byte will be taken will be 65 rather than 64 (Fig. 6.11).

We can now convert our flowchart into program coding, using symbol names (labels) in place of actual address numbers. Three labels, START, AGAIN, NXTONE have been used for addresses in the program memory, and four (COUNT, STOR, ADDR, SUM) for addresses outside the program memory. The curious appearance of

Example: The byte stored at address 100 (denary) is 64 (denary).

The byte stored at address 64 (denary) is 12 (denary).

The instruction: LDA,(100) will load *12* into the accumulator, since the instruction means: load the byte from the *address contained* at location 100.

If this is followed by INC(100) then the byte in address 100 is incremented to 65 (denary).

If the byte stored at address 65 is 6 (denary) then the instruction ADD,(100) will add 6 to the accumulator, since the instruction means: load the byte from the address contained at location 100.

Figure 6.11 *The indirect-addressing instructions of the (imaginary) instruction set (also used on SI-MPU-2).*

1.	START:	LDA 04H	;	put count number in
2.		STA,COUNT	;	store it in COUNT
3.		LDA,STOR	;	load in address byte of STOR
4.		STA,ADDR	;	put into ADDR for indirect loading
5.		LDA,(ADDR)	;	indirect load from STOR address
6.	AGAIN:	INC,(ADDR)	;	increment so that STOR=STOR + 1
7.		ADD,(ADDR)	;	indirect addressing used
8.		STA,SUM	;	store sum
9.		LDA,COUNT	;	get count number
10.		DEC	;	decrement count number
11.		JNZ,NXTONE	;	if not zero, loop back
12.		HLT	;	end of program
13.	NXTONE:	STA,COUNT	;	decremented number back to COUNT
14.		LDA,SUM	;	get sum in accumulator again
15.		JMP,AGAIN	;	return to loop
16.		HLT	;	end of program

Figure 6.12 *The loop-count program.*

some of the words chosen for labels is due to the limitation of a maximum of six letters which is imposed by many *assemblers* (the computer programs which convert these programs written in mnemonics and symbols into actual operating codes and data bytes in binary code).

The program (Fig. 6.12) looks more complicated than the flowchart, mainly because each step on a flowchart can require several microprocessor steps. The action of the program will be better understood when each step has been explained, keeping the action of the flowchart in mind. At the label START, which simply identifies the start of the program, the accumulator is loaded, using a load immediate, with the count number 04. This has then to be stored into the memory location labelled COUNT, so that it can be checked during each pass through the loop.

This done, the accumulator is now loaded with the number which is the address number, STOR, where the first of the five numbers is placed. So that the indirect addressing will operate correctly, this number now has to be loaded into another address, labelled ADDR, by the STA,ADDR step. These four steps set up the correct loop conditions so that the main part of the program can begin.

In step 5, the accumulator is loaded (indirect loading) from ADDR. This means that the address STOR is used, and the byte that is fetched into the accumulator is the byte from this STOR address, the first of the number to be added. Step 6 is labelled AGAIN, because this is an address we shall need to be able to identify, since it is the point to which the loop returns. At this step, the number held in address ADDR is incremented, increasing the amount from STOR to STOR+1, and so addressing the next number that is to be added in.

At step 7, the byte from STOR+1 is added into the accumulator, using the indirect addressing method, and step 8 stores the sum from the accumulator into the memory address SUM. This temporary stage is necessary because in this instruction set, the count number has to be fetched into the accumulator to be decremented and returned to memory. Some microprocessor types, notably the 6502, allow a number to be incremented or decremented at its memory address, without fetching it into the microprocessor. The fetch and decrement action is carried out in steps 9 to 12. Note that after the count number has been decremented, it must be put back into location COUNT, otherwise the unchanged number 04 will always be fetched from this address at step 9, and the program will never stop.

The test in step 12 is for a zero result, which will happen only when the number in COUNT was 1 and is decremented to zero. When this happens, the program moves to step 13, and halts, but until the last number has been fetched and added, the count number will not reach zero, and the section of program labelled NXTONE (step 14) will follow step 12. This action reloads the accumulator with the sum taken from memory, and then jumps for another pass through the loop to step 6, labelled AGAIN.

Now look at Fig. 6.13. This is an almost identical program, but instead of testing for zero accumulator content by using JNZ, this version uses JPZ — jump if the accumulator content is zero. In this way, the jump is taken only when the program is finished, and the normal run through the program is to steps 13 and 14, then back to AGAIN in step

[Other steps as previously...]

11.		JPZ,OUTIT	; out if count zero
12.		STA,COUNT	; decremented number back to COUNT
13.		LDA,SUM	; get sum back in accumulator again
14.		JMP,AGAIN	; return to loop
15.	OUTIT:	HLT	; end of program

Figure 6.13 *Amendments caused by substituting JPZ for JNZ. This makes the program one step shorter!*

1. LDA 04
2. STA,50
3. LDA 60
4. STA,100
5. LDA,(100)
6. INC,(100)
7. ADD,(100)
8. STA,55
9. LDA,50
10. DEC
11. JPZ,15
12. STA,50
13. LDA,55
14. JMP,06
15. HLT

Figure 6.14   *The program written for SI-MPU-2.*

6. This makes the program one step shorter by eliminating one of the HLT steps.

Finally, Fig. 6.14 shows this program written in its final form for the SI-MPU-2 simulator, with (denary) addresses and step numbers written in. This program should be tried, single stepping through the loop until the action is thoroughly understood.

Summary 6.3

An addition program can be used as an example of looping techniques in action. Five bytes can be added by using one LOAD followed by four ADD instructions, but an alternative is to use a loop which contains only one ADD instruction. An algorithm must be derived, and a suitable flowchart drawn up for the algorithm. Indirect addressing or a form of indexing will be needed to implement the loop, which is a counting loop, but one containing an operand which is altered on each pass through the loop.

Exercises 6.3

1. What type of program set makes repetitive actions easy to deal with?
2. What changes would be needed to the algorithm for loop adding if 50 numbers had to be added? What changes would be needed to a linear algorithm (LOAD, ADD, ADD, ADD, ...)?
3. What very important feature of loops does the answer to Q2 illustrate?
4. What would happen if the count number were not decremented on each pass?
5. What would happen if the address were not incremented on each pass?
6. Why is the address incremented and the count decremented?
7. Why must some form of indirect or indexed addressing be used to implement the program?
8. Why does the program look so much more complicated than the flowchart?
9. Why does the total have to be stored while the count is being decremented?
10. If a program of this type is faulty, what is the best method of testing it?

End-of-Chapter Test

1. State what is meant by a program loop.
2. Define (a) a holding loop (b) a counting loop (c) a searching loop.
3. Illustrate how each type of loop is used.
4. What type of loop can be used as a time delay?
5. Why should a time delay be useful?
6. What type of loop could be used in conjunction with a keyboard entry?
7. What step in a loop determines whether the loop is repeated or not?
8. What type of addressing method is needed in a searching loop?
9. Why do loop flowcharts sometimes have to be amended?
10. What is a bridging jump (or branch)?
11. How are memory addresses for jumps shown (a) on flowcharts (b) in a program?
12. Devise an algorithm for taking seven numbers from consecutive addresses in memory, subtracting 20H from each, and replacing them in their original addresses.

13. Draw a flowchart for the algorithm of Question 12.
14. Write a program, using the flowchart of Question 13, and using the instruction set shown in Fig. 6.8.
15. If, in a program devised from the flowchart of Question 14, each number was found to have had 20H correctly subtracted, but was placed in an address one greater than its original address, what fault would you look for in the program?

7

Buses and Interfaces

Buses

A BUS is a set of conducting lines to which the pins of the microprocessor and other chips can be connected. The reason for the name (the Latin omnibus means 'for all') is that several sets of chips can be connected to the same set of lines, and signals can be routed to and from different chips along the same set of lines. The three important buses of a microprocessor system are the data bus, the address bus, and the control bus.

Of these three, the address bus is used exclusively to send address information from the microprocessor to the other chips, mainly memory chips. The address bytes are generated within the microprocessor and are stored in the program counter and the store-address registers. These addresses are 'put on to the bus' by connecting the outputs of the registers to the address pins of the microprocessor, and this is an operation which is carried out by gates within the microprocessor itself, under the control of the microprogram.

The control bus signals vary from one microprocessor design to another, but some are outputs from the microprocessor, and others are inputs. One output that we have examined previously is the R/W control signal which allows the RAM memory to be either read or written, according to the signal voltage on this line. As an example of an input that is used for all designs, the interrupt (INT) line allows the normal action of the microprocessor to be interrupted at any time during a program by a signal on this line, which is an input control line.

Of the three buses, the data bus is the only one in which each line has to be capable of handling signals in each direction. Data signals must flow from microprocessor to memory during a write operation, and from memory to microprocessor during a read operation. Similarly when a port chip is connected to data lines, data bytes will be either read by the microprocessor from the port, or written by the microprocessor to the port.

A bus of this type is termed bidirectional, and it can only be used because of two features:

1. The microprocessor operates strictly in sequence, so that reading and writing can never occur at the same time.
2. The connections between chips and the data bus can be arranged so that only one input and one output is using the bus at a given time.

We have come across both of these points previously, but the second one now needs closer examination. A typical example is the use of a chip-enable pin on a memory chip, so that the chip is isolated from the data bus unless the enable pin is at the correct logic voltage. This type of switching between connection and isolation is known as tri-state control.

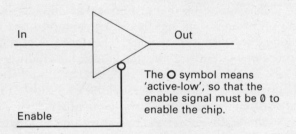

The O symbol means 'active-low', so that the enable signal must be 0 to enable the chip.

Figure 7.1 *A tri-state non-inverting buffer symbol.*

A typical tri-state controlled unit is shown in Fig. 7.1. The unit which is shown is termed a *non-inverting buffer*, and its action is described by the truth table. When the enable signal voltage is low, an input to the buffer causes an output of identical logic voltage, but when the enable signal voltage is high, the output 'floats', taking the same voltage as the bus line to which it is connected, and having no effect on the bus line. Fig. 7.2 shows the truth table for an inverting buffer. The only difference is that the output signal, when the buffer is enabled, is the inverse of the input signal.

In	Enable	Out
0	1	X
1	1	X
0	0	0
1	0	1

The symbol X means that the output can take either voltage; it will be at the voltage of the line to which it is connected, without affecting that voltage.

Figure 7.2 *The truth table for the buffer.*

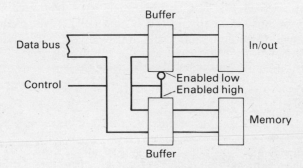

Figure 7.3 *Using tri-state buffering on data bus lines to separate memory chips from port chips. One buffer is enabled when the control line is low, the other when it is high.*

The tri-state control action can be applied to bi-directional lines, so that it is possible to use a single signal to control the connections of a complete data bus. In the circuit of Fig. 7.3, for example, a data bus is connected through two sets of bi-directional tri-state buffers to memory chips and to input/output terminals. The tri-state controls are activated by a signal from the microprocessor which ensures that for most of the time of a program, the microprocessor is connected to the memory, with the input/output terminals isolated. When the control signal (which could be obtained by decoding address lines) changes, probably as a result of a program instruction, the data bus is isolated temporarily from the memory, and connected instead to the input/output terminals. The scheme is not quite as simple as has been outlined here, because some method is needed to return the system to normal so that the next program instruction can be read from the memory, but similar principles are widely used.

A set of buffers which can send bus signals along long lines are, logically enough, known as bus drivers!

Summary 7.1

A bus is a set of conducting lines to which all the chips of a microprocessor system are connected. The three main buses are the data bus, the address bus and the control bus. Of these, the data bus is bidirectional, meaning that signals can pass along it in either direction. The address bus is used for address signals sent out from the microprocessor, and the control bus is used for a number of control signals, some outputs, some inputs. The bidirectional use of the data bus lines is possible only because the chips in the system have tri-state control — they can be isolated until an enable signal is received at the enable pin.

Exercises 7.1

1. What is a bus?
2. In what direction are bits transmitted along an address bus?
3. What feature distinguishes a data bus from the other buses?
4. How do address numbers originate?
5. Does a control bus consist of output lines only?
6. Can reading and writing ever be simultaneous?
7. What is meant by 'tri-state' control?
8. How is the tri-state pin of a chip controlled, and from where?
9. Write a truth table for a non-inverting tri-state buffer.
10. What signal determines the direction of signals on the data bus?

Interfacing

Interfacing means, for our purposes, the connection of a microprocessor system or part of a system, to other circuits, and it usually involves changes in the signals that are passed from one circuit to the other. The simplest type of interfacing is called buffering.

A buffer is a form of digital amplifier which is used to 'clean up' the shape of digital signals and also to provide the power that is needed to transmit signals along cables. The simplest buffer is the single non-inverting buffer, for which the output signal is of the same logic voltage as the input signal. If, however, the input signal has been degraded by the circuits through which it has passed (Fig. 7.4) with perhaps a logic 0

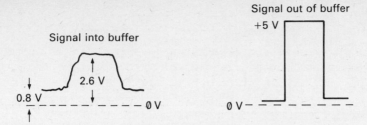

Figure 7.4   *The use of a buffer to 'clean-up' a signal.*

voltage that is about 0.7 V and a logic 1 voltage as low as 2.9 V, perhaps also a rather poor rise time from logic 0 to logic 1, then a non-inverting buffer output, with this sort of input, will be of the correct logic voltages, and with a shorter rise time.

The power-providing action is needed particularly when the signals of the microprocessor buses are to be used in other circuits. The microprocessor is a low-power device, and the extra capacitance of cables, plus the current needed to provide signals to other chips, may be too much of a load, causing degradation of the signals. In such cases, the connection of buffers between the microprocessor buses and the external circuits is highly desirable, because without such buffering, the signals may not meet the minimum specifications for logic voltages. As has been noted earlier, the buffering action is very often combined with tri-state control so as to be able to isolate sections of bus. The buffer may also play a part in the timing of the signals into and out of the microprocessor system.

A tri-state buffer has outputs which are isolated until its control pin is forced to the 'active' voltage, so that such a control, operated by the microprocessor, can ensure that an input, for example, can have no effect until the microprocessor is ready for it. This will be essential if the inputs and outputs of a system connect to the same set of lines.

For example, Fig. 7.5 shows a block diagram of a microprocessor system which uses a very primitive type of port for inputs and outputs. The port action is that there is a connection between the data bus and the eight port pins when the port-enable pin is at logic 0, and the port address number is selected by eight address lines. Now since the port must be bi-directional, we must ensure, by using tri-state buffer, that any outputs or inputs will not conflict with each other. This can be done by placing the tri-state buffers between the port outputs and the devices which will pass signals to the microprocessor or use signals from the

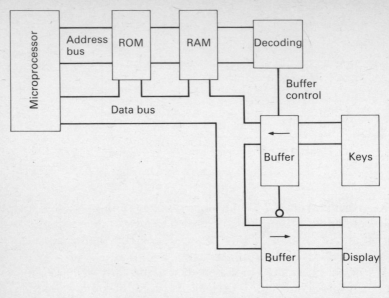

Figure 7.5 *A block diagram of a system which uses a simple port made from two sets of tri-state buffers.*

microprocessor. For example, a set of switches such as might be used for a keyboard can be connected to the port through a tri-state buffer. This buffer does not need to be bi-directional, because we can only pass signals from the switches to the microprocessor, never in the other direction. The tri-state buffer has its enable pin operated by a control voltage which is obtained by combining an address-decoding signal with the read control line of the microprocessor, so that the switches affect the data bus only when the correct address is placed on the address bus, and a read signal sent out, as would happen when a load ACC command from that address was programmed.

Similarly, a set of LED displays can be used to show the voltages on the data bus line, or to display these voltages in the form of a binary number. Since signals will only ever pass from the microprocessor to the LEDs, a one-way buffer will again be sufficient, and its enable pin will be operated by a combination of address-decoded signal and microprocessor write signal. The interface in this case would not be a simple buffer but a *latch*, which will also store the data. The reason is that the buffer would connect the LEDs to the bus for only a microsecond or so, and the display would not be visible. By using a latch, the binary number is stored, and the LEDs remain lit until another signal is sent to the latch.

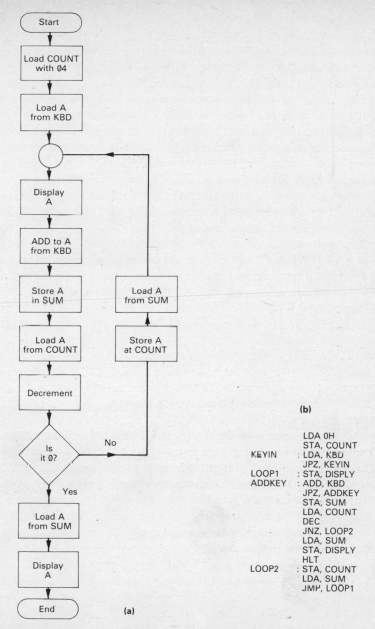

(b)

```
                     LDA 0H
                     STA, COUNT
         KEYIN    : LDA, KBD
                     JPZ, KEYIN
         LOOP1   : STA, DISPLY
         ADDKEY : ADD, KBD
                     JPZ, ADDKEY
                     STA, SUM
                     LDA, COUNT
                     DEC
                     JNZ, LOOP2
                     LDA, SUM
                     STA, DISPLY
                     HLT
         LOOP2   : STA, COUNT
                     LDA, SUM
                     JMP, LOOP1
```

(a)

Figure 7.6 *A flowchart* **(a)** *and a program* **(b)** *for the addition of five numbers input from a keyboard. This assumes that a latch is used to keep figures displayed after a STA,DISPLY command. A waiting loop has been used in the keyboard input.*

The program of Fig. 6.11 can be modified, for example, so that instead of an indirect memory load at each pass through the loop, the microprocessor enables the keyboard and holds in a waiting loop until a key is pressed, so inputting a number. When all five numbers have been entered in this way, the program can then be made to activate a display, and put the contents of the memory address SUM on to the data lines.This can be done in a loop if no latch is used, so that the display is visible. This illustrates another very useful application of waiting loops, to overcome the problems caused by the brief time that the microprocessor takes to execute an instruction. If the microprocessor activates a keyboard for only 3 microseconds, no human operator is ever likely to produce an input. Similarly, if the result of the sum is displayed for only 3 microseconds, no human eye will see it, and a waiting loop is a software alternative to a latch chip in each case.

The flowchart and the form of the program that will be needed is shown in Fig. 7.6. This is, incidentally, an excellent program to try out on the EMMA microprocessor assessment unit, when the 6502 microprocessor codes are mastered, or on the MENTA unit, using Z–80 mnemonics.

Summary 7.2

Interfacing means the connection of one system to another, so that signals can be passed from one system to the other. Interfacing microprocessor systems may require changes in signal voltages, methods of coding, line voltages, or the number of lines used for signals. A buffer is a simple form of interface which maintains the form of logic signals used in microprocessor systems. The buffer will improve the shape of each pulse and will permit more current to be drawn from its output than the microprocessor system could supply. A buffer which has tri-state control is particularly useful.

Exercises 7.2

1. What is interfacing?
2. Suggest a type of interface device which could fit between a microprocessor system and an electric motor (switching on and off).
3. What is a buffer?
4. How can a buffer be said to 'clean up' a signal?
5. What is the difference between an inverting and a non-inverting buffer?
6. What additional feature is very useful for a buffer?
7. How can buffers be used for input/output?
8. Draw a flowchart for the steps needed to obtain an input from a buffer placed between a keyboard and a microprocessor system.
9. Why is a waiting loop needed for key-inputs and visual outputs?
10. What would be an essential feature of a buffer on the data bus between the microprocessor and the memory?

Code Conversion

Another aspect of interfacing which is very often needed is code conversion. We may, for example, have a memory full of characters in ASCII code and we want to transmit them to a printer, such as an IBM Selectric, or a Murray-code terminal, which uses a different form of code for each character. An interface between such units must then

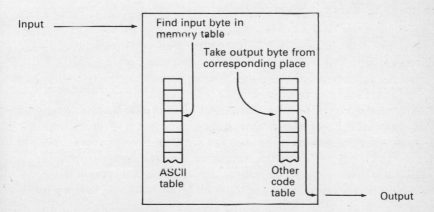

Figure 7.7 *Using a 'look-up table' method for code conversion. This can often be done using a ROM in which the data inputs are used as addresses which cause output bytes in the other code to be read.*

carry out the task of converting each ASCII byte into the corresponding byte in the other code, and changes of voltage may be needed also. The usual technique is to use another microprocessor system, programmed with a set of *look-up tables*. The details of the technique are beyond the scope of this book, but briefly, one block of memory is loaded with the full set of ASCII codes in sequence, and another block with a set of the other codes for the same characters in order. When a byte is read into the system, it is compared with each byte in the ASCII set, and each comparison is counted. When a match is found, the number of places along from the start of the block is used to locate a byte which is the same number of places from the start of the other block, where the other code is stored. This byte is then fed out as the output of the interface.

Voltage Conversion

An interface may have to carry out conversions between different logic voltage standards. One very common application is converting normal logic voltages (called TTL voltage levels) in which logic 0 is 0.8 V or less, and logic 1 is 2.4 V or more) into the system called RS232, which uses, for example, +12 V for logic 1 and −12 V for logic 0 (the voltages need not be exact, providing that equal positive and negative voltages are used). The RS232 system is widely used for printers and for terminals, which are a combination of keyboard and VDU.

Timing Conversion

Microprocessor systems which use eight lines of data bus will deal with one complete byte of data at each stage in a program. Transmitting one byte at a time to remote units, such as terminals and printers, needs at least eight lines — data bus and some control lines — and this is reasonably convenient when only short connections are used. This type of connection is often known as a parallel connection, so that a parallel printer, for example, will use this type of multiple connection. The alternative is serial connection, in which each bit of each byte is sent separately along a pair of lines, using a coding method which allows the receiving device to re-combine the bits into a byte again.

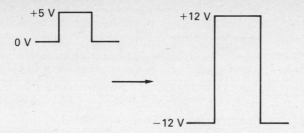

Figure 7.8 *Voltage conversion. In this illustration a TTL pulse is converted to RS232 voltage conventions.*

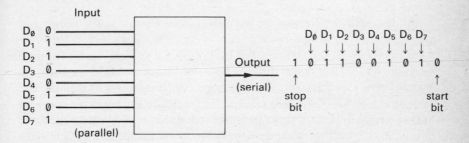

Figure 7.9 *Timing conversion — in this example, parallel bits on eight lines are converted to a serial stream (first bit of the stream shown on the right) with a start bit and a stop bit. Two stop bits are sometimes used.*

One much used pattern of transmission and reception of serial data is laid down by the RS232 standard, and it involves sending a start bit, the eight bits of data, and a stop bit for each byte transmitted. The time between each pair of bits is rigorously controlled so that the receiver can keep exactly in step with the transmitter. An interface for such purposes is a fairly complex circuit, requiring a quartz-crystal clock circuit to ensure correct timing. The timing is measured in terms of the *Baud rate*, meaning the number of changes of voltage on the line per second. Baud rates from 110 to 1200 are fairly common, and faster rates can be used. At the higher rates, a control signal, called a 'handshake', is used to make sure that the receiver is not overloaded. If the receiver cannot accept more signals (perhaps because of a slow mechanical action, like paper feed), the handshake signal on a third line will stop transmission until the voltage on the handshake line changes.

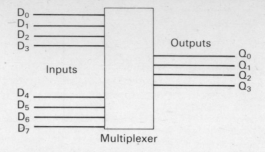

Figure 7.10 *A multiplexer, which connects its output lines alternately to two sets of inputs. Corresponding de-multiplexers (decoders) can be used to reverse the process.*

Finally, an interface may be used to cause a change in the number of lines that are used to convey signals. Fig. 7.10 shows a dual four-line multiplexer. When the control pin is at one logic voltage, the set of lines marked A0 to A3 is connected to the output lines Q0–Q3 but when the control pin is at the other logic voltage, the lines marked A4–A8 are connected to Q0–Q3. This allows the voltages on eight lines to be transmitted along only four lines, in sequence, and by extending the principle any number of lines of data can be sampled by the eight lines of a data bus. Providing that the line voltages do not change in less than the time needed to carry out at least one complete set of samples, this multiplexing action is a very satisfactory way of putting information from large numbers of lines into a microprocessor system.

Summary 7.3

Another form of interface is code conversion, in which bytes sent to or from the microprocessor system are converted into different bytes. Unless the conversion is simple (such as add or subtract a constant), this is most easily done by using a 'look-up table', comparing each byte with a set in memory and finding the corresponding byte from another set.

Voltage conversions are often needed when signals are sent along lines, so as to use standard voltage values. Timing conversions are often needed for line transmission also, because single-line transmission allows only one bit at a time to be sent or received. Conversion to different numbers of lines are also used.

Exercises 7.3

1. Why is 'code conversion' needed?
2. What is a look-up table?
3. When is a look-up table *not* needed for code conversion?
4. What are the normal logic voltages of a microprocessor system?
5. What is meant by RS232?
6. In what devices are RS232 signals used?
7. What is a parallel connection?
8. What is a serial connection?
9. Why are ten bits transmitted for each byte along a serial line?
10. Why is a quartz crystal circuit used in a serial interface?

End-of-Chapter Test

1. A chip is described as a 'bus driver'. What is it intended to do?
2. List the three important buses of a microprocessor system.
3. An 8-bit microprocessor system has a 16-line bus. Which bus is this?
4. Which of the microprocessor buses is bidirectional?
5. A microprocessor uses an interrupt input. On what bus will this signal be placed?
6. List the two features which make the use of a bidirectional bus possible.
7. What is meant by a 'floating' output?
8. What is an interface?
9. What is a buffer?
10. Why should a buffer use tri-state control?
11. Why are buffers placed between the microprocessor and other chips?
12. Sketch a block diagram for a microprocessor system with input and output through buffers.
13. Write an algorithm and draw a flowchart for a program which inputs numbers from a keyboard and displays a running total.
14. What is the purpose of a look-up table?
15. Why are (a) voltage conversion (b) timing conversion needed for some interfaces?

Appendix A

SI-MPU-2 SIMULATOR

```
1 REM SI-MPU-2 PROGRAM TO SIMULATE MPU ACTION FOR TEC MICROPROCESSOR SY
STEMS (2), PUBLISHED BY HOLT-SAUNDERS - COPYRIGHT.
2 DATA "LDA","STA","ADD","DEC","HLT","JMP","CMP","JPZ","JNZ","INC"
3 REM BY IAN SINCLAIR, FOLLOWING AN IDEA BY E. PARR
4 REM WRITTEN IN SINGLE-STATEMENT LINES
5 CLS:REM USE SCREEN CLEAR
7 CLEAR 1000
8 DIM P$(99)
10 PRINTTAB(10)"SI-M.P.U."
20 PRINT
30 PRINT"PLEASE CHOOSE BY TYPING NUMBER FOLLOWED BY 'RETURN'
40 PRINT
50 PRINTTAB(5)"1. WRITE ASSEMBLER PROGRAM."
60 PRINTTAB(5)"2. RECORD ASSEMBLER PROGRAM."
70 PRINTTAB(5)"3. REVIEW ASSEMBLER PROGRAM."
80 PRINTTAB(5)"4. REPLAY RECORDED PROGRAM."
90 PRINTTAB(5)"5. RUN PROGRAM."
100 PRINTTAB(5)"6.END."
110 INPUT I
120 IF I>0AND I<7 THEN GOTO 170
130 PRINT"INCORRECT SELECTION- NUMBER MUST BE BETWEEN 1 AND 6
140 FOR N=1TO 1000
150 NEXT N
155 CLS:REM USE YOUR OWN SCREEN-CLEAR COMMAND
160 GOTO 30
170 ON I GOTO 500,1000,1500,2000,2500,3000:REM MACHINES WITHOUT THIS CO
MMAND WILL HAVE TO USE LINES SUCH AS 170 IF I=1THEN500 171 IF I=2THEN10
00 ETC.
500 CLS:REM USE OWN SCREEN-CLEAR
510 FOR N=1TO99:REM THIS IS WRITE SECTION
512 P$(N)=""
514 NEXT N
515 REM CLEAR VARIABLES FOR NEXT WRITE
520 PRINT"ADDRESS NUMBER (DENARY) IS SHOWN AT EACH STEP."
530 PRINT"USE STANDARD NOTATION (SEE TEXT), WHICH IS:"
540 PRINT"    LDA 09"
550 PRINT"    LDA;15"
555 PRINT"    lda;(100)"
560 PRINT"    25 OR 04 (MUST BE 2 DIGITS)"
570 PRINT"REMEMBER SPACES AND SEMICOLONS"
```

114

```
580 PRINT"TYPE X   TO TERMINATE ENTRY"
590 PRINT"PRESS RETURN TO PROCEED"
600 INPUT Z$
610 CLS:REM USE OWN SCREEN-CLEAR
620 N=1
630 PRINT "MEMORY ADDRESS";N;:REM KEEP PRINT IN SAME LINE
640 INPUT P$ :REM INSTRUCTION
642 IF LEN(P$)>3 THEN V=VAL(MID$(P$,5,3))
643 IF LEN(P$)<=3 THEN V=VAL(P$)
644 IF V<256 AND V>=0 THEN 650
646 PRINT "INCORRECT NUMBER SIZE - MUST LIE BETWEEN 0 AND 255"
648 GOTO 630
650 IF P$="X" THEN GOTO 690
660 P$(N)=P$
670 N=N+1
680 GOTO630
690 N=N-1
700 PRINT"ENTRY FINISHED"
710 PRINT"PRESS RETURN TO GO TO MENU."
720 INPUT Z$
725 CLS :REM CLEAR SCREEN
730 GOTO30
1000 REM NEED A SUITABLE RECORDING ROUTINE FOR YOUR OWN MACHINE HERE.
1400 PRINT"NO RECORDING ROUTINE-PRESS RETURN TO GO TO MENU"
1410 INPUT Z$
1415 CLS:REM CLEAR SCREEN
1420 GOTO 30
1500 CLS:REM CLEAR SCREEN - REVIEW PROGRAM
1510 PRINT"TO REVIEW YOUR PROGRAM-"
1520 PRINT"PRESS RETURN FOR EACH STEP."
1530 PRINT"PRESS 1,RETURN TO ALTER STEP."
1540 FOR J=1TON
1550 PRINT J;" ";P$(J)
1560 INPUT Z
1570 IF Z=1 THEN GOTO 1660
1580 NEXT J
1590 PRINT "END OF PROGRAM."
1600 PRINT "INPUT 0 TO EDIT, 1 TO RETURN."
1610 INPUT Z
1620 IF Z=0 THEN GOTO 1660
1630 IF Z=1 THEN CLS
1635 GOTO 30
1640 PRINT"INCORRECT SELECTION"
1650 GOTO 1600
1660 PRINT "INPUT NUMBER OF INSTRUCTION."
1670 INPUT Z
1680 PRINT Z;" "; P$(Z)
1690 PRINT "INPUT NEW INSTRUCTION"
1700 INPUT P$
1710 P$(Z)=P$
1720 PRINT "ANY MORE? 1 FOR YES, 0 FOR NO."
1730 INPUT Z
1740 IF Z=1 THEN 1660
1745 CLS
1750 GOTO 30
2000 REM NEED SUITABLE REPLAY ROUTINE FOR YOUR OWN MACHINE HERE
2400 PRINT"NO REPLAY ROUTINE-PRESS RETURN TO GO TO MENU."
2410 INPUT Z$
2415 CLS
2420 GOTO 30
2500 CLS:REM SCREEN CLEAR - RUN PROGRAM
2505 F=0
2510 PRINT"TO RUN NORMALLY, TYPE 1."
2520 PRINT"TO SINGLE-STEP, TYPE 2."
2530 PRINT "THEN PRESS RETURN."
2540 INPUT Z
2545 G=0
2550 IF Z=2 THEN G=255
2555 J=0
```

```
2560 J=J+1
2565 IF J>N THEN 2600
2570 GOSUB 5000:REM PARSING AND SELECTION
2580 IF G=255 GOTO 2680:REM ANALYSIS SECTION
2590 GOTO2560
2600 PRINT"END OF PROGRAM"
2610 PRINT"TO SEE THE QUANTITIES IN MEMORY"
2620 PRINT"INPUT STEP (MEMORY) NUMBERS."
2630 PRINT"USE 0 TO RETURN TO MENU."
2640 INPUT Z
2650 IF Z<>0 THEN GOTO 2660
2652 CLS
2654 GOTO 30
2660 PRINT "CONTENT OF  MEMORY";Z;" IS "P$(Z)
2670 GOTO 2610
2680 PRINT "STEP ";J
2690 PRINT "CONTENT OF MEMORY IS ";P$(J)
2700 PRINT "ACCUMULATOR CONTAINS ";A
2710 PRINT "PRESS RETURN TO PROCEED"
2720 INPUT Z$
2730 GOTO2590
3000 END
5000 L$=LEFT$(P$(J),3):REM PARSE COMMAND
5005 IF VAL(L$)<>0 THEN RETURN
5010 IF MID$(P$(J),4,1)=";"THEN F=255
5015 IF MID$(P$(J),5,1)<>"(" THEN 5020:REM TEST FOR INDIRECT
5016 F=100
5017 R$=MID$(P$(J),6,2):REM GET INDIRECT ADDRESS
5018 GOTO 5025
5020 R$=RIGHT$(P$(J),2)
5025 V=VAL(R$)
5027 RESTORE
5030 FOR X=1TO10
5040 READ Q$(X)
5050 IF Q$(X)=L$ THEN GOTO 5090
5060 NEXT X
5070 PRINT "FAULTY INSTRUCTION"
5080 CLS
5085 GOTO 30
5090 ON X GOTO 6100,6200,6300,6400,6500,6600,6700,6800,6900,7000:REM US
E IF X=1 STATEMENTS IN ABSENCE OF ON X GOTO
5100 F=0
5110 B=0
5120 IF A>255 THEN A=A-256
5140 RETURN
6100 IF F=255 THEN GOTO 6130:REM LDA
6110 A=V
6112 IF F<>100 THEN 5120
6113 AA=VAL(P$(V)):REM INDIRECT VALUE
6114 A=VAL(P$(AA))
6115 GOTO 5120
6120 GOTO 5100
6130 A=VAL(P$(V))
6140 GOTO 5100
6200 P$(V)=STR$(A):REM STA
6210 GOTO 5100
6300 IF F=255 THEN GOTO 6330:REM ADD
6301 IF F<>100 THEN 6310
6302 AA=VAL(P$(V)):REM INDIRECT ADD
6303 A=A+VAL(P$(AA))
6304 GOTO 5100
6310 A=A+V
6320 GOTO 5100
6330 A=A+VAL(P$(V))
6350 GOTO 5100
6400 A = A-1 : REM DEC
6410 GOTO 5100
6500 GOTO 2600 : REM HLT
6600 J=V-1:REM JMP
```

```
6610 GOTO 2590
6700 B=0 :REM CMP , B IS FLAG
6710 IF F=255 THEN GOTO 6740
6720 B=A - V
6730 GOTO 5120
6740 B=A-VAL(P$(V))
6750 GOTO 5120
6800 IF B<>0 THEN GOTO 5100 :REM JPZ
6810 J=V-1
6820 GOTO 2590
6900 IF B=0 THEN GOTO 5100:REM JNZ
6910 J=V-1
6920 GOTO 2590
7000 IF F=100 THEN 7030
7010 A=A+1 : REM INC
7020 GOTO 5100
7030 P$(V)=STR$(1 + VAL(P$(V))):REM INDIRECT INC
7040 GOTO 5100
7100 END
```

Appendix B

The Menta Assessment Unit

MENTA is a small British designed and built Z-80-based assessment unit which has no built-in display system. By plugging the unit into an ordinary domestic television receiver, the contents of the RAM memory of MENTA can be displayed on the television screen, with 256 bytes, in hex, shown at any one time. The total RAM of 1K can therefore be shown in four 'pages' of display. This allows the whole of a short program, or indeed, several short programs, to be displayed in a glance, and is very valuable for reviewing programs. A very considerable advantage over other units is the fact that this type of display highlights the first byte of each instruction, so that operator and operands are clearly separated.

This improved output method is matched by the input. Most assessment units use hex entry of address and data only, but MENTA, in addition to allowing hex entry, allows the entry of assembler language commands directly from the keyboard. Some 94 per cent of all Z-80 commands can be entered in this way (the remainder, which are seldom-used functions, can be entered as hex codes), so that for educational purposes, MENTA is a miniature assembler, but with the enormous advantage of low price and small bulk. Cassette recording and replay of the whole of the RAM is provided for.

Programs entered by the use of assembler language can be edited, and it is not necessary to calculate addresses or displacements, as these can be inserted automatically by the unit, making use of an 'address store'

118

facility. It is also possible, for the first time in a low-cost unit of this type, to insert new code in the middle of a program without having to re-enter the whole of the following code.

MENTA has so many advantages for the teacher of Microelectronic systems work that it should be seriously considered for any course of this type. Port outputs/inputs are provided for connection to external circuits, and the manual contains a full listing of Z-80 codes.

The agents are:
>Dataman Designs,
>Lombard House,
>Dorchester,
>Dorset DT1 1RX

Price at the time of writing was £115 plus VAT (if applicable).

Index

Accumulator 18, 26
Active state 105
Addition 8, 16, 69, 92
Address decoding 57
Address store 27
Addressing methods 74
Algorithm 13, 33, 40, 43
ALU 25, 30
AND-gate 20, 70
Arithmetic and Logic instructions 66, 69
Arranging numbers 33
ASCII codes 28, 73, 109
Assembler, MENTA, 118

Baud rate 111
Bidirectional bus 102
Binary addition 8
Binary numbers 1
Binary subtraction 12
Bit 2
Bootstrap monitor 48
Branching flowchart 37
Branching program 81
Branch instruction 72
Branch step 86
Buffer 104
Bug program 48
Bus 101
Bus drivers 103
Byte 2

CALL instruction 72
Carriage-return byte 83
Carry bit 9, 28, 72
Chip select/enable 58
Clock cycle 28
Clock generator 28
Code conversion 109
Compare instruction 70
Complement 11
Conditional branch/jump 72
Control, read/write 61
Control and timing 25
Converting binary to hex 5
    binary to octal 7
    denary to hex 6
    denary to octal 7
    hex to binary 5
    hex to denary 6
    octal to binary 7
    octal to denary 7
    signed binary to denary 12
Corruption, program, 63
Counting loop 81, 82
Crash, program, 63

Data bus 18, 102
Data transfer instruction 66
Decision step 37
Decoders 60
Degradation of signal 105
Demultiplexers 60

Deriving flowchart 40
Digital amplifier 104
Displacement byte 77
Division 13
Don't care values 46
Dynamic RAM 32

Effective-address (EA) 74
Enable, chip, 58
Excess-3 code 1
Execute cycle 29
Extended addressing 76
Extended block diagram 25

Fast idiot 34
Fetch 29
Fetch-execute cycle 29
Flag register 9, 20, 27
Flags 28, 72
Flip-flops 16
Floating output 102
Flowchart 34
    addition 42, 45, 94
    branching 37
    derivation 40
    iterative 37
    linear 37
    looping 37
    rearrangement 88

Garbage 63
Gate 16
Gate, AND 20, 70
    OR 21, 70
    XOR 21, 70
Gating 20, 59
Gray code 1

Handshake signal 111
Hexadecimal (hex) 4
Hexadecimal scale 4
High-level language 48
Holding loop 81

IBM Selectric code 109
Immediate addressing 75
Indexed addressing 77
Indirect addressing 78
Input/output symbol 34
Instruction groups 66

Instruction register 26, 46
Instruction set 66
    SI-MPU 93
INT line 101
Interfacing 104
Iterative flowchart 37

Jump instruction 72

Keyboard input 107, 108

Labels 40, 95
Latch 106
Linear flowchart 37
Load 17, 68
Logic instruction 70
Logical manipulations 20
Look-up tables 110
Looping flowchart 37, 88
Looping program 81, 92, 96

Machine control 49
Memory 16, 52
Memory-mapped port 68
Memory organization 54
MENTA assembler 118
Microprogram 46
Microprocessor system 49
Monitor program (bug) 48
Multiple-jump steps 89
Multiplexer 112
Multiplication 13
Murray code 109

Negative numbers 10
Nibble 54
Non-inverting buffer 102
Number cruncher 14
Numbers, binary 1
    negative 10

Octal scale 4
Operand 24
Operator 25
OR-gate action 21, 70
Over-written byte 18

Parallel connection 110
Pass of loop 81
PC-relative addressing 77

Port chips 68, 105
Process symbol 36
Program corruption 63
Program counter (PC) 26
Program crash 63

RAM 49, 52
Random selection 53
Reading 18
Read/write (R/W) 29
    control 61, 101
Refresh memory 55
Register 16
    accumulator 26
    address-store 27
    instruction 26, 46
    program counter 26
    status 9, 20, 22, 28
Register-indirect addressing 78
Register-to-register transfers 67
RET instruction 72
ROM 48
Rotating 20, 70
Rotation 23
RS-232 110–111

Searching loop 81, 83
Select, chip 58
Selecting memory units 53
Serial connection 110
Shift 22, 70
Shifting 20
Sign bit 28, 72
Significance 1

SI-MPU-2 simulator 93, 114
Sorting algorithm 33, 35
Static RAM 52
Status register 9, 20, 27
    bits 28, 72
Storage 16
Store 17, 68
Store-address register 27, 46
Subroutine 73
Subtraction, binary 12, 69
Subtrahend 12
Symbolic names 40
Symbols, flowchart 36

Test-and-branch instruction 66, 72
Timing 28
Trace table 46, 47
Tri-state control 102
Truth table 20
Two-byte addition 43
Two's complement 11

Unconditional branch/jump 72

Volatile memory 62
Voltage conversion 110

Writing 18

XOR gate 21, 70

Zero bit 28, 72
Zero-page addressing 77